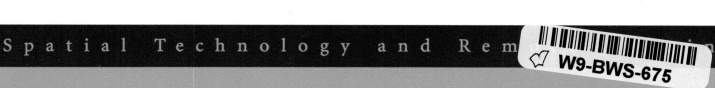

S p a t i a l   T e c h n o l o g y   a n d   R e m o t e   S e n s i n g

Introduction to

Geographic Information Systems

Tools and Processes

Statement of Copyright

Introduction to GIS Tools and Processes

Table of Contents - Student Manual

Preface vii
How will I use this book? vii
Features in this book viii
About ESRI's ArcGIS viii
Teacher Materials ix
About the Authors x

Unit 1: Understanding Spatial Data

An overview into the two main ArcGIS Programs: ArcMap and ArcCatalog and an explanation of data types used throughout the ArcGIS Suite. Unit Features a brief narrative of each technology and an exercise applying the concepts. The final lesson is an exercise combining all areas.

Lesson 1   Introduction to ArcMap 3
Lesson 2   NavigatingAcrMap 25
Lesson 3   Introduction to ArcCatalog 31
Lesson 4   Navigating ArCatalog 67

Unit 2: Managing a Data Inventory

An introduction into the basic processes and skills essential for geospatial technicians taught in the framework of homeland security applications. Each lesson features detailed narratives of each phase, an exercise applying the concepts, and a lesson review to reinforce key concepts.

Lesson 1   Displaying Geospatial Data 73
Lesson 2   Managing Geospatial Data 101
Lesson 3   Creating Geospatial Data 123
Lesson 4   Analyzing Geospatial Data 145
Lesson 5   Preparing Geospatial Data 171
Lesson 6   Planning& Building a Local Data Inventory 203

Glossary:

Glossary G-1

Table of Contents

Preface

This hands-on book is is an ideal tool for anyone interested in ArcGIS. It introduces you to two very important components of the ArcGIS program; ArcCatalog and ArcMap. You will be provided with step by step instructions that will take you from learning the basics of these programs; like launching a map, viewing and editing metadata to creating new shapefiles and eventually to building a local data inventory. While learning these valuable skills, you will be using the same geospatial tools that industry calls upon to find solutions.

Lessons are built around applications to show process beyond "buttonology". This type of learning promotes knowledge of tools as well as building a solid foundation for users to make good decisions when faced with choices that will ultimately affect the end users, customers, and colleagues who will benefit from analysis. The skills that that are taught in this book allow the user to go directly from classroom application to real world application. Geospatial technology is enhancing all manners of life and industry, and we want you to have a working knowledge of these tools.

This book is designed for secondary, post secondary, and professionals. There are neither prerequisites nor is previous GIS experience necessary. We provide all directions and data needed to complete the directed tasks

How will I use this book?
This book is designed to be completed sequentially with lessons that provide an overview of ArcGIS with subsequent lessons examining introductory processes in ArcGIS in the context of Homeland Security Applications. Each lesson will focus on a necessary concept that is fundamental to a users growth as geospatial technician.

Features in this book:

- **Step-by-Step Instructions** –an easy to follow format relevant for novice to experienced ArcGIS users
- **Lesson Content** – provides a narrative overview of an important topic relevant to the fundamental concepts of geospatial technology
- **Lesson Exercise** – applies lesson content to an exercise that features "real world" tasks at work
- **Lesson Review** – reinforces knowledge gained with exercises to identify key terms and concepts
- **"Knowledge Knugget" boxes** – boxes found in the margin that features tools, tips and tricks that may enhance your experience with geospatial technologies
- **Finished Layouts** – to provide a self assessment tool to ensure successful completion of each lesson

About ESRI's ArcGIS

ArcGIS is a software suite developed by Environmental Services Research Institute, Inc. (ESRI) designed to analyze and model geospatial data. Of the software suite in ArcGIS, this book will use three major components; ArcMap, ArcToolbox and ArcCatalog. ArcMap is the primary part of the suite that will be used throughout the book to display, create, and analyze different types of geospatial data. ArcToolbox contains various geoprocessing tools used throughout the ArcGIS suite to complete various tasks such as creating buffers, merging shapefiles, and address locators. ArcCatalog is the "virtual filing cabinet" where users create, manipulate, or preview data and metadata. ArcGIS is widely accepted and used among today's GIS professionals and students. Using this software will make for a smooth transition for the student to take their GIS skills from the classroom to the workplace or other academic pursuits.

Teacher Materials

The STARS Introduction to Geographic Information Systems and Remote Sensing Concepts is designed to be used in a variety of learning environments. For classroom environments, a teacher's edition is available with the following enhancements:

Overviews – Each lesson in the teacher's edition comes with a lesson overview page for instructors with boxes in the margins that provide quick reference to lesson goals. The "What Will You Teach" section provides the instructor with a bulleted list that includes a list of goals for the lesson GIS skills learned. The "How Will You Teach It" section provides the instructor with the procedures to introduce the topic and have the students complete the lesson. The initial introduction provides a unique perspective on the topic that is covered in the lesson that will enhance teacher's abilities to facilitate learning in a classroom environment. The overview also describes in detail the skills taught in the lesson as well as additional information that may be necessary to complete the lesson. Lesson related links for further study are also supplied with each overview page.

PowerPoint Presentation Notes – Each student lesson comes with a PowerPoint presentation that provides an overview of the lesson including concepts, the skills they will cover, and the study area involved. The teacher's manual is supplemented with detailed descriptions and commentary for each slide allowing a diverse range of instructors to lead classroom lecture. For every presentation, the student manuals will have notes sheets, complete with thumbnail pictures of the slides from the presentation, with lines for notes beside them.

Assessments – Lessons will conclude with a full page color layout of a successfully completed exercise. If questions are presented within a lesson, the teacher's manual includes answers to those questions.

About the Authors

Eddie Hanebuth is founder and president of Digital Quest, a Mississippi-based development and training-oriented company that produces GIS instructional material for educational institutions. He chairs the U.S. Department of Labor's National Standard Geospatial Apprenticeship Program and the SkillsUSA Geospatial Competition Committee, and runs the SPACESTARS teacher-training laboratory in the Center of Geospatial Excellence, NASA's John C. Stennis Space Center.

Liz Rotzler has six years experience in geospatial technology education. After teaching GIS in the classroom using the STARS curriculum and certification, she has spent the past three years working in development of GIS/RS curriculum. She has co-authored and edited the first book in Digital Quest's aGIS series, Introduction to Geospatial Technologies as well as worked in the popular Digital Quest SPACESTARS series.

Austin Smith has been part of the Digital Quest team for four years. He is currently the Vice President of Development and Support and also serves as the chair of the S.T.A.R.S. Geospatial Certification Committee. He has experience in information technology development, implementation, and training in a variety of public and private organizations. At Digital Quest he has co-authored or edited over 30 titles. With Digital Quest's STARS series, he serves in authoring, planning, and final editing.

Understanding Spatial Data

Learning Topics

Introduction to ArcMap

Introduction to ArcCatalog

Introduction to
Geographic Information Systems
Tools and Processes

Unit One

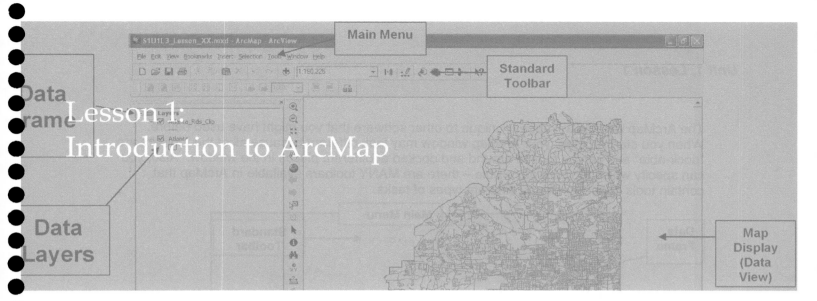

Lesson 1: Introduction to ArcMap

Think about where you are sitting or standing right now. There are a number of different ways that you can describe not only where you are – in a room, in a building, in a city, in a county or parish, in a state, and so on; but also there are a number of other ways of describing the area you currently occupy – 70% humidity level, 80° in temperature, and more. For just this one point on the Earth, there's a lot going on. Every point on the Earth has a tremendous amount of data associated with it and thanks to geospatial technology, any or all of this data, can be shown on a map. The ability to create maps with various types of data is extremely important to us as we make decisions everyday that affect our lives. Geospatial technology is a decision-making tool and geospatial (GIS) software is the vehicle to that decision-making power.

ArcGIS

In this course, you will be introduced to skills and concepts that are central to the understanding of geospatial technology using an industry-standard software suite – ArcGIS. ArcGIS is created by Environmental Systems Research Institute (ESRI), a leader in the GIS field. ArcGIS is not the only software program that ESRI has created, but it is the most widely used GIS programs on the market and it provides analysis capabilities that can be applied to nearly every scenario that can be analyzed using geospatial technology.

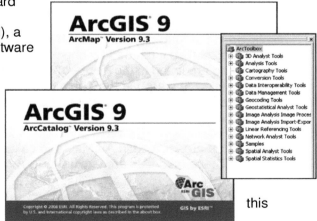

ArcGIS is known as a "suite" of software and it contains three software components that will be used in this course – ArcMap, ArcCatalog, and ArcToolbox. You will explore each of these in this unit and in various parts of this manual.

ArcMap

ArcMap is the software component of ArcGIS that allows you to display and perform analysis on spatial data. In addition, you can create map layouts using ArcMap. Throughout this course, you will spend the majority of your time using this program.

The ArcMap window will appear unique to other software that you might have used before. When you start ArcMap, your ArcMap window may look different because all of the toolbars are "dock-able" and can be moved around and docked at different places in the window. Also, you can specify which toolbars are visible – there are MANY toolbars available in ArcMap that contain tools to perform many different types of tasks.

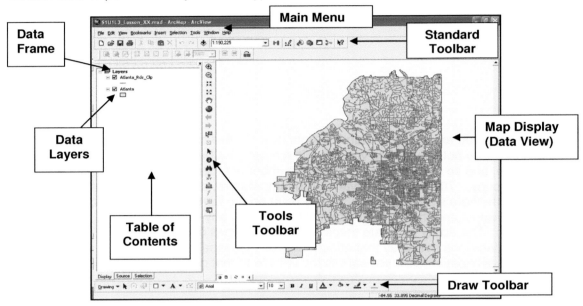

The individual data files that are used in ArcMap are called **data layers**. Data layers can be added and removed as needed. They can also be "checked" ☑ so that the contents are displayed in the map display or "unchecked" ☐ so that they are not visible in the map (but can easily be displayed if needed by checking them). These layers are organized in **data frames** in ArcMap and are listed in the **Table of Contents**. In this example, there is only one data frame (currently titled "Layers") in the Table of Contents. With ArcMap, you can have more than one data frame at a time. This gives you the ability to create layouts with multiple maps being displayed at once.

The files that are created in ArcMap and called **map documents** and will have a file extension of **.mxd**. These mxd files are exclusive to ArcMap.

Spatial Data in ArcMap

ArcMap provides many different options and tools for displaying spatial data. Spatial data is displayed as **raster** or **vector**. Raster data is data that is made up of equally sized individual cells representing a certain area on the Earth. Two examples of raster data include **images** files and **grid** files. For images, each cell has a certain "spectral signature" associated with it that is determined by the type of land cover that is found in that area. Depending on the type of data that is organized as a grid data layer, each cell contained in a grid has a value associated

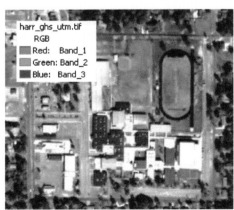

with it – the value may be an elevation or an average temperature a land use code.

Most of the data that you will use in your studies is considered **vector** data. Vector data is data that represents geographic "features" on the Earth and symbolizes these features using point, line, and polygon shapes.

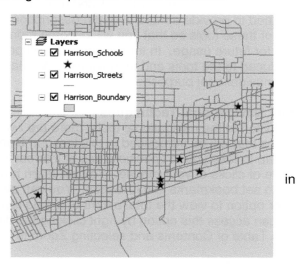

For the most part, vector data does not represent continuous data. Vector data represents geographic features. Unlike a raster data set, if you click on different areas the same geographic feature, you get the same value. There is no variation in values that you have with raster data. Thus, vector data is considered as "homogeneous."

in

ArcMap Tools
The ArcMap tools that will be introduced in this lesson are shown in the table below.

Tool Icon	Tool Name
🔍⊕	Zoom In
🔍⊖	Zoom Out
⌐⌐	Fixed Zoom In
⌐⌐	Fixed Zoom Out
⚫	Full Extent
ⓘ	Identify
✋	Pan
⬅	Previous Extent
➡	Next Extent
▨	Select Features
◪	Clear Selected Features

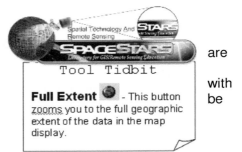

In the lesson documentation, the purpose of each of these tools is included in a Tool Tidbit box. These boxes are designed to provide a more in depth description of the uses of these tools so that you will become more familiar with them for future usage. In this lesson, you will primarily be introduced to the tools and techniques in ArcMap that have navigational purposes.

Other Navigational Techniques

In addition to the navigational tools, there are two other navigational options that will be introduced in this lesson. **Bookmarks** are used in ArcMap to take you to specific areas of interest in the map display with a couple of clicks of your mouse. You will learn how to use bookmarks and how to set bookmarks in this lesson. You can also use the **Zoom to Layer** option to view the total extent of a data layer in the map display. You can access this option by right clicking the data layer of interest in the Table of Contents and selecting Zoom to Layer.

Displaying Data

As you zoom to a layer, the perspective of the data you are looking at will most likely change. The spatial coordinates for the data you are viewing are listed at the bottom right corner of the ArcMap window. Depending on the coordinate system being used, these coordinates may be displayed using degrees minutes seconds, decimal degrees, meters, miles, or some other unit of measure. The coordinate system can be edited to best suit your mapping needs. The information on how to edit the coordinate system and when it should be done will be covered in a later unit.

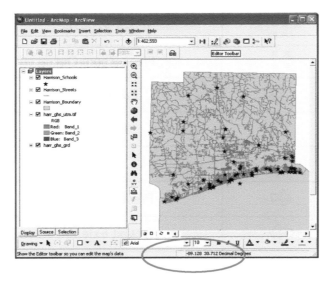

Conclusion

The ArcGIS software suite is made up of a group of software programs designed to display, manage, and analyze geospatial data. ArcMap is a one of the programs within the suite that you will most frequently use in this course. There are many different navigational tools in and techniques in ArcMap that can be used to view spatial data. Raster data, such as images and grid files, and vector data, such as points, lines, and polygons, can be viewed in ArcMap simultaneously.

Lesson 1: Introduction to ArcGIS – ArcMap

By now, you understand that spatial data is information about geographic features – that is, data about things that have a specific place on the surface of the Earth. You can view spatial data by looking at maps. However, paper maps are static – they do not change based on which geographic features are of interest to you. Geographic information systems (GIS) offer a dynamic interface to view and analyze geographic data. ArcGIS is a software suite designed to create and manage GIS projects. ArcGIS is a product of Environmental Systems Research Institute (ESRI), a Redlands, CA-based company and makers of industry-standard geospatial technology software products.

ArcGIS is a suite of 3 programs: ArcMap, ArcCatalog and ArcToolbox. In addition, ModelBuilder allows you to create your own custom geoprocessing tasks and ArcGlobe works with 3D Analyst extension program to allow you to visualize three-dimensional data in exciting, dramatic ways.

This unit is designed to introduce you to the two (2) frequently used ArcGIS programs – ArcMap and ArcCatalog and give you experience navigating in each program. You will be introduced to terminology relevant to the software programs and will prepare for the skills and tools that you will learn in future units.

Launching ArcMap:

1. **Start ArcMap** by **double-clicking** the ArcMap shortcut on your computer desktop (or by clicking the Windows Start button **start**, **pointing** your mouse to **Programs/ArcGIS** and **selecting** **ArcMap**.)

2. When the ArcMap window appears, make sure that the radio button next to the (Start using ArcMap with) ⊙ An existing map: option is selected.

At the bottom of the ArcMap startup dialog box, make sure that  is highlighted.

Click OK.

Navigate to **C:\STARS\GIS_RS_Tools** and ***select*** the **S1U1L1_Lesson.mxd** file.

Click Open.

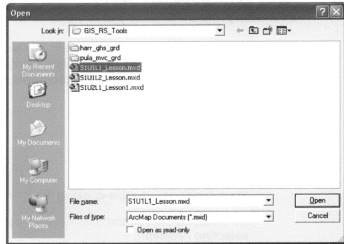

The **map document** will open and display data layers for Little Rock (and Pulaski county), AR. The map view will be zoomed in on an air photo of a portion of Little Rock.

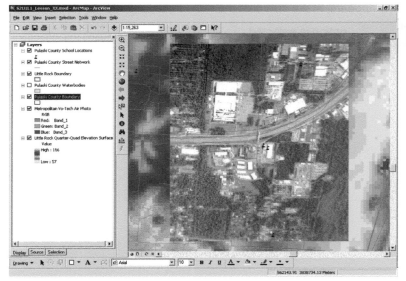

Terminology Tip

The files that you create when you work on a project in ArcMap are called ***Map Documents***. The file extension for an ArcMap map document is ***.mxd***.

As you look at the map display, orient yourself just as you would if you were looking at a paper map. As you move up the map display, geographic orientation moves north. As you move to the right of the map display, geographic orientation moves east and so on.

3. To save this ArcMap map document in your student folder, *select* **Save As…** from the **File** menu. **Name** the new file **S1U1L1_Lesson_XX** (where **XX** is your initials) and **save** the file in your student folder.

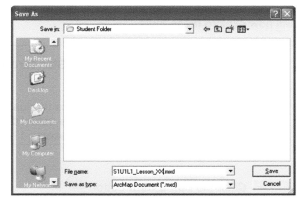

Take a minute to refer to the different parts of the ArcMap window. Make a note of where the toolbars and other components of the ArcMap window are located.

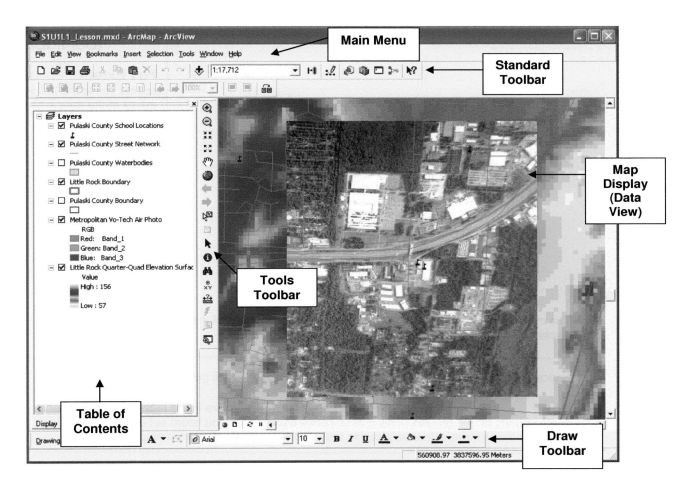

The **Table of Contents** is the area of the ArcMap window that lists the data layers that are used in this GIS project. The individual data layers are listed in the **Table of Contents** as they are layered in the map display. The **Table of Contents** can be used like a legend to read the data contained in the map display.

Navigating a Map Display:
ArcMap has several navigational tools that allow you to view the data in the map display as needed. You will now explore these tools.

> Before you move on, look at the **Standard Toolbar** at the top of your **ArcMap** window and notice
>
> the **Map Scale** box 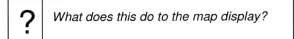 (due to screen size and resolution, your number
> may differ from this number). This manual scale box provides the scale of the map display in
> terms of how it compares with the actual distance on the ground. In this example, one (1) unit of
> measure in the map display represents 17,712 units on the ground.

4.   ***Click*** the **Full Extent** button ⚫.

> **?** *What does this do to the map display?*

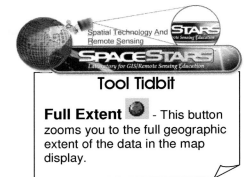

Tool Tidbit

Full Extent ⚫ - This button zooms you to the full geographic extent of the data in the map display.

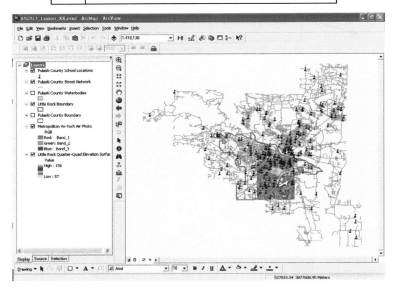

The map display zooms out to the full geographic extent of all of the data contained in the map display.

> Notice how the map scale changed when you zoomed to the full geographic extent of the map
>
> display `1:444,034`. Now, one (1) unit of measure in the map display represents
> 444,034 units on the ground. This makes sense because, as you zoom out in the map display,
> you are viewing more actual geographic area on the ground in the same map display area.

Notice in the **Table of Contents** that the data layers contained in this project that are drawn in the map display. This is noted in the **Table of Contents** by the **checkbox** to the left of the layer name ☑. Only one (2) layers – Pulaski County Waterbodies and Pulaski County Boundary – are not checked ☐. These layers are included in the ArcMap map document but the features contained in the layer are not currently drawn in the map display. You can toggle a data layer on and off in the map display by checking or unchecking the layer's checkbox.

5.   ***Click*** to **check** both the **Pulaski County Waterbodies** and the **Pulaski County Boundary** layers **checkboxes** ☑ in the **Table of Contents** to draw the layers features in the map display. The features contained in this data layer will appear in the map display.

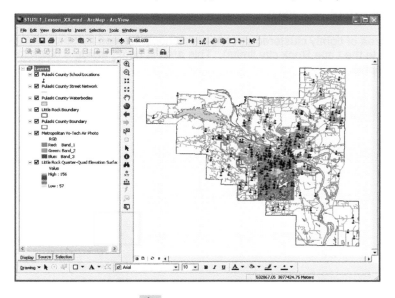

6.   ***Click*** the **Previous Extent** button .

> **?** *What does this do to the map display?*
> *What did it do to map scale?*

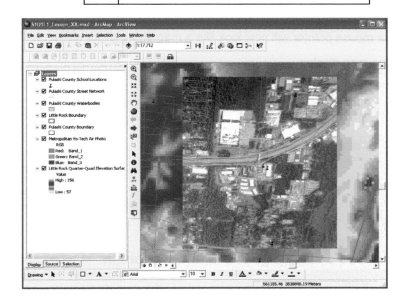

Tool Tidbit

Previous Extent ⬅ - This button zooms you to the last extent that you were viewing in the map display.

Next Extent ⬅ - Once you have used **Previous Extent**, you can use this button to move forward in extent settings.

The **Previous Extent** navigational tool  allows you to go back to the last extent that you were using in the map display. Once you use the **Previous Extent** button, you can use the **Next Extent** button to move forward through the extent settings. You can think of these buttons somewhat like the **Back** and **Forward** buttons on your Internet browser – the only difference is that you are not going to a different page; you are viewing the same data at a different extent.

7. ***Zoom*** to **Full Extent** again.

Notice the area of tightly clustered streets and schools located in the central portion of Pulaski County (inside the Little Rock city limits). This area is noted inside a black rectangle in the following graphic.

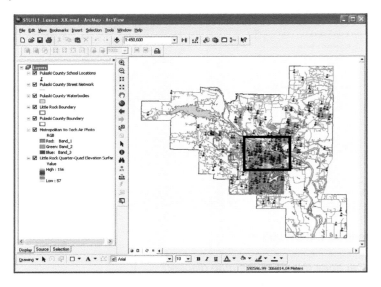

In order to get a closer look at this area on the map display, you can zoom in to the area.

8. ***Click*** the **Zoom In** tool . Your mouse pointer will now look like a magnifying glass with a plus sign when you move it over the map display.

9. ***Click and drag*** a box over the area identified in the graphic above to zoom in to the area of interest.

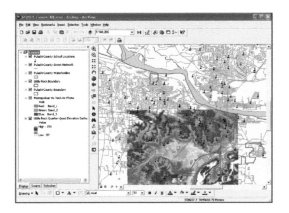

Tool Tidbit

Zoom In - This button zooms you to an area of interest in the map display. You can either click on an area or click and drag to get a closer look at an area.

When data is added to ArcMap, the program somewhat organizes the layers by points, lines, then polygons so that the layers can be easily seen in the map display. However, ArcMap does not organize the data layers within their feature types. For instance, one polygon layer that is above another polygon layer in the Table of Contents may actually obscure some of the polygon features in the other layer. In those cases, you may have to rearrange the order of the data layers in the Table of Contents. This is the case with the Pulaski County Waterbodies and Little Rock Boundary layers.

In the area that you are currently viewing in the map display, you may notice that part of the Little Rock boundary can be seen close to the large waterbody that is north in the map display. In order to make sure that the waterbody polygons do not hide part of the boundary of Little Rock in the map display, you must reorder these layers in the Table of Contents.

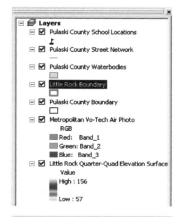

10.   In the **Table of Contents**, *click* the **Little Rock Boundary** data layer and *drag* it above the **Pulaski County Waterbodies** layer.

Notice that the city boundary can now be seen along the waterbody in the map display. The city boundary polygon feature is not obscured by the waterbody polygon because the city boundary layer is now layered over the waterbody layer in both the Table of Contents and the map display.

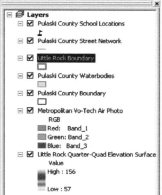

Tool Tidbit

Identify ⓘ - This button displays information (attributes) about individual features when the features are clicked on in the map display.

Identifying Features:

To find out more about the features in the map display, you can use the **Identify** tool ⓘ to view information about features.

11.   *Click* the **Identify** tool ⓘ . The **Identify Results** dialog box will appear.

From the **Layers** drop down list, *select* **Pulaski County Waterbodies**.

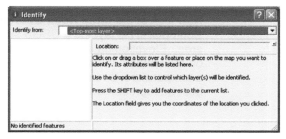

Using your mouse, *click* on the **large waterbody** in your map display.

Information about this geographic feature will display in the **Identify Results** dialog box. This waterbody feature is the **Arkansas River**.

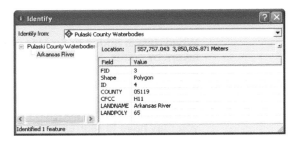

12. **Change** the **Layer** to identify to **Pulaski County School Locations** in the **Identify Results** dialog box.

    Click one of the schools in the map display to view information about the feature.

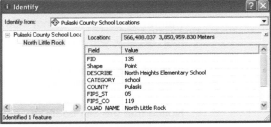

13. **Change** the **Layer** to identify to **Pulaski County Street Network** in the **Identify Results** dialog box.

    Click one of the street segments in the map display to view information about the feature.

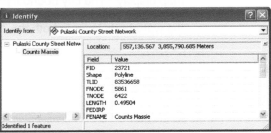

14. **Change** the **Layer** to identify to **Pulaski County Waterbodies** in the **Identify Results** dialog box again.

    Click on the **Arkansas River** feature again.

    Take your mouse and **click** in a different area of the **Arkansas River** than you just clicked.

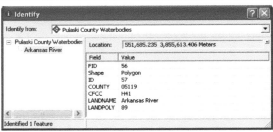

> **?**   *Did the information differ?*

Regardless of where you click in a single geographic feature of a vector (feature) data layer, the "value" or information about that feature is the same. Vector data is known as "homogeneous" data because values within individual features are the same throughout the feature.

Now look at the **Little Rock Quarter-Quad Elevation Surface grid data layer** in the Table of Contents. This layer is a USGS surface model that displays elevation data over a geographic area. Elevation data is displayed as a grid because elevation is "continuous data." Everywhere on the surface of the Earth has an elevation value associated with it. Likewise, all locations have a temperature, a rainfall value, etc. All of these data types are considered "continuous" and can best be represented using grid data layers.

Notice in the Table of Contents that the elevation values for the surface model in this project range from a high of 156 to a low of 57. The elevation values of areas within the grid are represented with the range of colors seen in the Table of Contents. Each individual cell in the grid has an elevation value associated with it. This particular USGS elevation grid has a cell size of 30 which means that one cell in the grid represents a 30 X 30 meter area on the surface of the Earth.

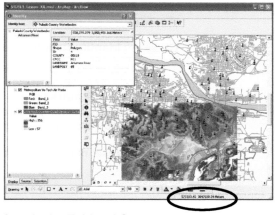

Note: *The reason why grid "resolution" is measured in meters is that all USGS elevation surfaces use the Universal Transverse Mercator (UTM) coordinate system that uses meters as its unit of measure. As you move your mouse across the map display, notice that the location coordinates corresponding to the position of your mouse display at the bottom of the ArcMap window. Because the USGS grid uses UTM coordinate system, this project uses this spatial reference. If the Geographic Coordinate System were used for this project, the coordinates would be displayed in degrees of longitude and latitude.*

15. ***Right click*** the **Little Rock Quarter-Quad Elevation Surface** in the **Table of Contents** and ***select*** **Zoom to Layer**. You will now see the entire extent of the elevation grid in the map display.

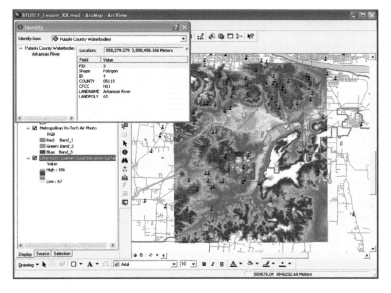

16. ***Change*** the **Layer** to identify to **Little Rock Quarter-Quad Elevation Surface** in the **Identify Results** dialog box.

Click somewhere on the **elevation grid**. Notice the information in the **Identify Results** dialog box.

The pixel value will be given in the dialog box. The pixel value reflects the elevation value of the corresponding location on the Earth.

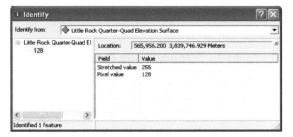

---

17.   *Click* somewhere else on the **elevation grid**. Notice the information in the **Identify Results** dialog box.

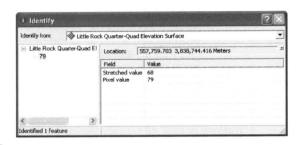

The pixel value will probably be different. This is Because grid data is continuous and, just as different Locations on the Earth have different elevations, the cells in this grid have different pixel values reflecting elevation.

18.   *Close* the **Identify Results** dialog box by *clicking* the ☒ in the upper right hand corner of the dialog box.

Using & Setting Bookmarks:

Another interesting navigational technique in ArcMap is the use of bookmarks to set certain geographic extents within the ArcMap map document.

19.   To use a preset map bookmark, *select* **Bookmarks, Lake Maumelle** from the **Main** menu.

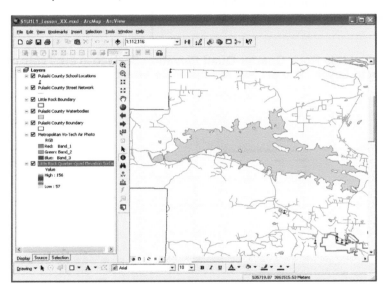

20.   If you would like more experience with the **Identify** tool, use this tool to identify the waterbody in your map display.

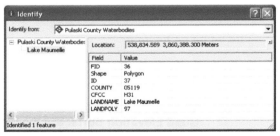

21.   *Zoom* to **Full Extent** ⬤.

22.   *Zoom in* 🔍 on a different waterbody in Pulaski County. Use the **Identify** tool ⓘ to determine the name of the waterbody feature.

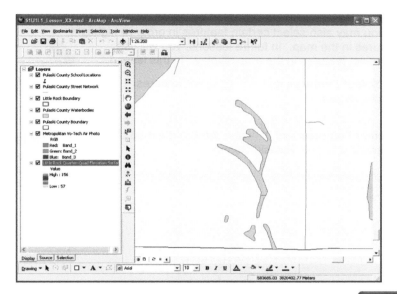

23. **Select** **Bookmarks, Create…** from the **Main** menu. The **Spatial Bookmark** dialog box will appear.

24. **Name** the **bookmark** with the name of the waterbody that you have chosen and that is shown in your map display.

Click  **OK** . The bookmark is now set.

25. **Click** to return to the **Previous Extent** 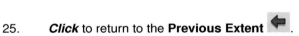 .

26. **Select** **Bookmarks, Your Waterbody Name** from the **Main** menu. You should return to the spatial bookmark that you just set.

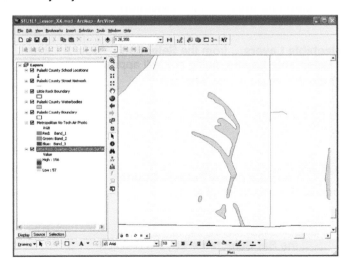

27. **Zoom** to **Full Extent** 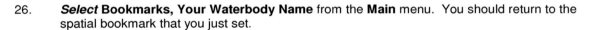 .

Selecting Features:
In the map display, you may also select certain features in order to highlight them from other features in the map. In future lessons, you will use other techniques to select features in the map display. In this lesson,

you will just use the **Select Features** tool  to select features interactively in the map display.

Tool Tidbit

Select Features  - This button allows you to highlight features in the map display by clicking on them.

28.   ***Click*** the **Select Features** tool and ***click*** in the map display on the northernmost school in Pulaski County.

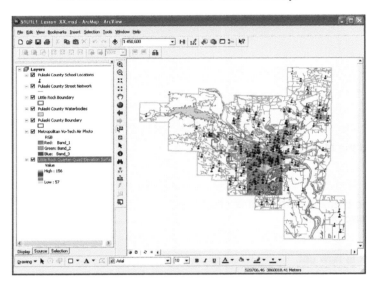

When you do this, you will notice that the school, the surrounding streets and the county boundary are all selected. This is caused by all data layers being specified as "selectable layers." To remedy this, you can set only the school layer as the selectable layer.

29.   ***Choose*** **Clear Selected Features** from the **Selection** menu.

30.   ***Choose*** **Set Selectable Layers…** from the **Selection** menu. The **Set Selectable Layers** dialog box will appear. All layers in the map document will be selected ☑.

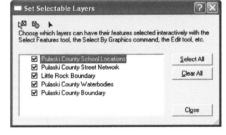

Click to [Clear All] and then ***check*** **Pulaski County School Locations** as the only selectable layer.

Click to [Close] the **Set Selectable Layers** dialog box.

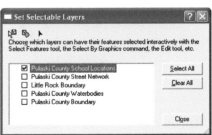

31. Use the **Select Features** tool 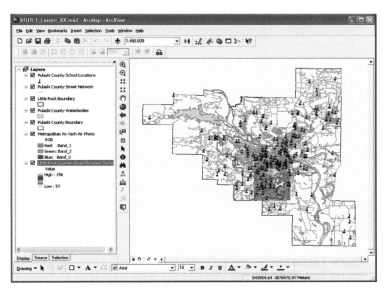 to select the northernmost school again. This time only the school should be selected.

32. ***Choose*** **Zoom to Selected Features** from the **Selection** menu. You will zoom to the school that you just selected.

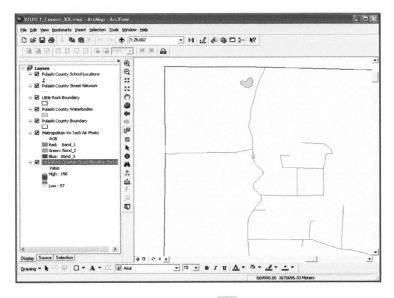

33. ***Click*** the **Fixed Zoom Out** tool to zoom out slightly in the map display. ***Click*** **Fixed Zoom Out** several more times until the map scale is approximately 1:150,000.

If necessary, ***use*** the **Fixed Zoom In** tool to zoom back in slightly.

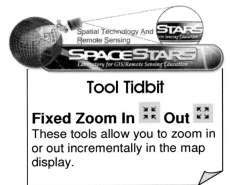

Tool Tidbit

Fixed Zoom In **Out**
These tools allow you to zoom in or out incrementally in the map display.

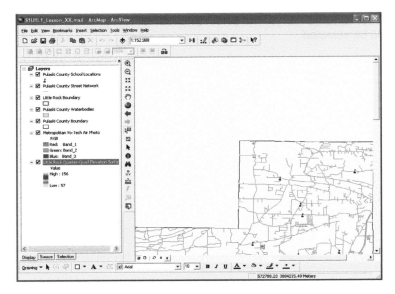

Viewing Layer Attribute Tables:

It is important to understand that, even though you see the data displayed as features in the map display, the data is organized in tables known as attribute tables. Whereas the feature on the map is the spatial entity, the table contains the information about the entities, or the feature's attributes.

34. **Right click** the **Pulaski County School Locations** layer in the **Table of Contents** and **select Open Attribute Table**. The attribute table for this layer will open.

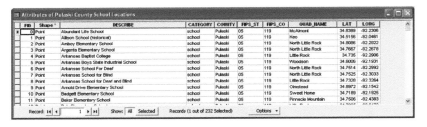

Each school feature in the map display corresponds to a record in this table.

Click to show only Selected features. The school that you selected in the map display will be the only record shown in the table.

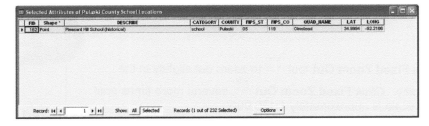

The selected school was **Pleasant Hill School**, a historical school that no longer functions as a school building.

Close the attribute table by *clicking* the  in the upper right hand corner of the table.

35.   *Click* the **Pan** tool .

Click and *drag* with the **panning mouse pointer** to move back toward the **Little Rock** area in the map display.

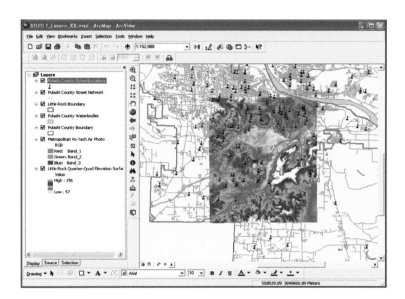

Tool Tidbit

Pan - This tool allows you to click and drag to move around the map display.

36.   At the bottom of the map display, *click* to switch to **Layout View** . In **Layout View**, the map that you create is displayed in page layout. In future lessons, you will add additional map elements to the layout page and print the map.

37.   *Save* the map document.

38.   To exit ArcMap, *select* **Exit** from the **File** menu **OR** *click* the in the upper right hand corner of the ArcMap window.

Lesson 1: Introduction to ArcGIS – ArcMap Lesson Review

Key Terms
Use the lesson or index provided in the back of the book to define each of the following terms.

1.   ArcGIS

2.   ArcMap

3.   Data layers

4.   Data frames

5.   Raster data

6.   Vector data

7.   Bookmarks

Global Concepts
Use the information from the lesson to answer the following questions. Use complete sentences for your answers.

8.   What type of file is created in ArcMap? What extension will it have?

9.   Describe two ways to quickly navigate to a certain set extent or layer?

10.   How does raster data differ from vector data?

ArcMap Tools

Place the letter of the source next to the correct description. Each will be used only once.

Tool Icon

Use

_____11.

_____12.

_____13.

_____14.

_____15.

_____16.

_____17.

_____18.

_____19.

_____20.

_____21.

A. Zooms out from the layout page

B. Zooms in to the layout page

C. Allows you to zoom out in a fixed amount

D. Allows you to zoom in a fixed amount

E. Allows you to view the entire geographic extent of the layers

F. Allows you to view attributes of a selected feature

G. Allows you to select feature(s) in the display

H. Allows you to return to the previous extent

I. Allows you to return to the next extent

J. Allows you to reposition your view within the data frame

K. Allows you to clear any selected features

Let's Talk About It...

Answer the following question and share the responses with your instructor and classmates.

22.   Why is vector data said to be homogeneous? Provide some examples of vector data.

In the last lesson, you were introduced to the ArcGIS suite of geospatial analysis software. The focus of that lesson was on navigating the ArcMap program. Remember that ArcMap is the program that allows you to display and analyze spatial data and create map layouts.

In this lesson, you will use the navigational and other skills that you learned in Lesson 1 to answer a series of questions. If necessary, refer to Lesson 1 for detailed information on ArcMap tools. Record your answers on the S2U1L2 Lesson Worksheet that accompanies this lesson in this manual.

1.   **Start ArcMap** by **double-clicking** the ArcMap shortcut on your computer desktop (or by clicking

the Windows Start button **start**, **pointing** your mouse to **Programs/ArcGIS** and **selecting** ArcMap.)

2.   When the ArcMap window appears, make sure that the radio

button next to the (Start using ArcMap with) ⊙ An existing map: option is selected.

At the bottom of the ArcMap startup dialog box, make sure that Browse for maps... is highlighted.

Click OK .

Navigate to **C:\STARS\GIS_RS_Tools** and **select** the **S1U1L2_Lesson.mxd** file.

Click Open .

The **map document** will open and display data layers for Gulfport (and Harrison county), MS. The map view will be zoomed in on an air photo of a portion of Gulfport.

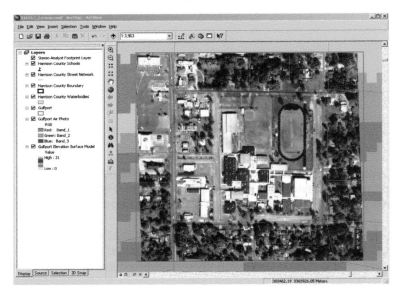

| ? | 1. What features do you recognize in the air photo? **Give answer on lesson worksheet**. |

3. To save this ArcMap map document in your student folder, *select* **Save As...** from the **File** menu. **Name** the new file **S1U1L2_Lesson_XX** (where **XX** is your initials) and **save** the file in your student folder.

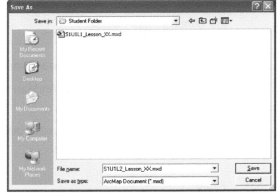

Look at the **Table of Contents** for this map document. Familiarize yourself with the data layers included in this project.

| ? | 2. Name three feature data layers used in this map document – a point, a line and a polygon.   **Give answer on lesson worksheet**. |

Notice the map scale for the map display.

| ? | 3. List the current map scale for the map display before you change the extent. **Give answer on lesson worksheet**. |

4.   *Zoom* to **Full Extent**.

> **?** 4. List the map scale for the map display at full extent. **Give answer on lesson worksheet**.

5.   *Go to* **Bookmark 1** in the map display.

> **?** 5. What is the name of the river that is shown at this bookmark? **Give answer on lesson worksheet**.

6.   *Zoom* back to the **Previous Extent**.

> **?** 6. What extent does the map display show now? **Give answer on lesson worksheet**.

7.   *Go to* **Bookmark 2** in the map display.

> **?** 7. What school is shown at this bookmark? **Give answer on lesson worksheet**.

8.   In the **Table of Contents**, notice the elevation range for the **Gulfport Surface Model**.

> **?** 8. What is the elevation range for the Gulfport elevation grid shown in the map display? **Give answer on lesson worksheet**.

9.   *Fixed zoom out* until you see another school to the east and another school to the west.

> **?** 9. What is the name of the school to the west when you fixed zoom out? **Give answer on lesson worksheet**.

10.   *Zoom* to the **Gulfport Air Photo** layer.

> **?** 10. What is the name of the school shown in the air photo? **Give answer on lesson worksheet**.

11.   *Save* the map document.

12.   *Exit* **ArcMap**.

Lesson 2: Navigating ArcMap

Name: _____

As you are prompted in your lesson activity, answer the following questions regarding ArcMap tools and navigational procedures that you perform in the lesson exercises.

1.   **What features do you recognize in the air photo?**

2.   **Name three feature data layers used in this map document – a point, a line and a polygon.**

3.   **List the current map scale for the map display.**

4.   **List the map scale for the map display at full extent.**

5.   **What is the name of the river shown at Bookmark 1?**

6. **What extent does the map display show when you go to the Previous Extent?**

7. **What school is shown at Bookmark 2?**

8. **What is the elevation range for the Gulfport elevation grid shown in the map display?**

9. **What is the name of the school to the east when you fixed zoom out?**

10. **What is the name of the school shown in the air photo?**

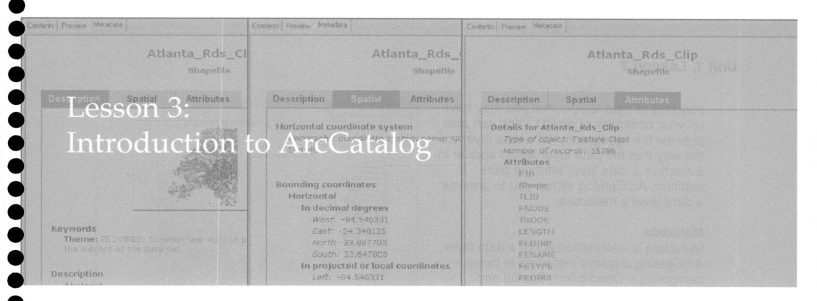

ArcMap is a program that allows you to view and analyze data. Behind the scenes there is a program that will allow you to organize and manage the many files that are involved in a GIS program. In this lesson, you will be introduced to that program, ArcCatalog.

ArcCatalog provides a wealth of capabilities in terms of data management. This lesson will give you experience navigating ArcCatalog and learning more about its capabilities. In future lessons, you will utilize ArcCatalog to manage existing data, create new data and manage/edit data properties.

ArcCatalog

ArcCatalog is the ArcGIS program that allows you to preview data, read metadata, create new data layers and manage data. The ArcCatalog window looks similar to the ArcMap window in many ways; the catalog tree on the left appears like the table of contents, the display window on the right looks like the map display window (you can even view data and previously created maps there), the menus at the top are similar as well. The main difference is that no analysis can be performed in ArcCatalog.

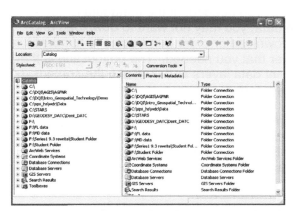 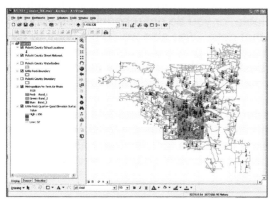

ArcCatalog (left) has some similarities in appearance to ArcMap (right). The main difference between the two is that no analysis can take place in ArcCatalog.

Spatial Data in ArcCatalog

ArcCatalog allows you to preview data before you use it. Adding data layers to ArcMap for the sole purpose of previewing it or trying to determine what data you need to use would take a considerable amount of extra time. ArcCatalog allows you to preview the

contents of data folders that you may have on your computer or on CD. You can also preview the "geography" of a data layer (or the way that the data layer would appear in a map) or a data layer attribute table. In addition, ArcCatalog allows you to preview a data layer's metadata.

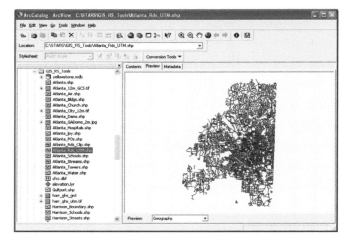

Metadata

Metadata is information about a data layer. ArcCatalog displays metadata in three categories – Description, Spatial and Attribute. Metadata includes information about who collected the data, the purpose of the data, the attributes that are included in the data layer, the coordinate system that the data layer uses, contact information regarding how to get the data, and other information about the data.

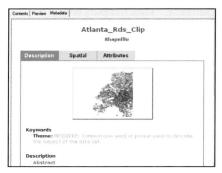

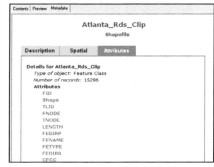

When you are looking for data for a project online, it is a good habit to view the metadata associated with different data layers to determine what you needed. If the metadata is not available, or you do not recognize the source of the data, you may not want to download and use that data. You also would not want to have to download or order hundreds and hundreds of data layers to preview to determine the one data layer that you needed. The bottom line is that it is best to read the metadata associated with different data layers and then use the contact information in the metadata to acquire the data that you need.

In order to ensure that metadata uses a standard format, the Federal Geographic Data Committee (FGDC) has established a set of metadata standards that should be used when creating metadata for data that has been collected. By using these standards, it simplifies the sharing of geographic data to individuals, businesses and organizations that use geospatial data everyday.

Metadata Styles

There are different metadata style sheets that are used to display metadata. The ones displayed below are FGDC ESRI, FGDC Classic, and FGDC FAQ (Frequently Asked Questions). These are just three examples of how metadata can be viewed in ArcCatalog. The style sheet you use to read metadata depends solely on your preference.

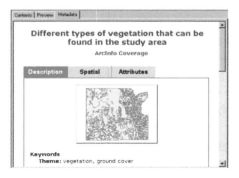

FGDC ESRI style

FGDC FAQ style

FDGC Classic Style

ArcCatalog Capabilities

As mentioned earlier in this lesson, about the only thing that cannot be done in ArcCatalog is analysis. ArcCatalog does provide the capability to preview data. It can also allow a user to view metadata as well as create and edit metadata. In ArcCatalog, you can copy and paste, delete and rename files with ease. Many geographic datasets often consist of several files instead of one file (like a word document does). When renaming or moving these files outside of ArcCatalog, meticulous attention should be paid so that all parts of that dataset are taken care of. A typical street file dataset, for example, has seven files associated with it! If you want to rename that street file, you have to rename all seven separately outside of ArcCatalog. If you are in ArcCatalog, it will show as one file and renaming that one file would automatically rename any associating files also.

ArcCatalog Tools

The ArcCatalog tools that will be introduced in this lesson are shown in the table below.

Tool Icon	Tool Name
	Details
	Large Icons
	List
	Thumbnails
	Edit Metadata
	Create Thumbnail
	Search
	Launch ArcMap

As you work through the lesson activity for this lesson, the purpose of each of these tools is included in a Tool Tidbit box. These tools, which primarily deal with navigating in ArcCatalog, will be used again at various times throughout this course.

Conclusion

The ArcGIS software suite is made up of a group of software programs designed to display, manage and analyze geospatial data. ArcCatalog is a program that allows you manage geospatial data in numerous ways - from reorganizing files, to viewing metadata, to previewing files. About the only thing that you cannot do in ArcCatalog is analyze data. There are many different navigational tools that can be used to preview spatial data in ArcCatalog. Before using a dataset in a project, it is always a good idea to preview the file's metadata. ArcCatalog gives you the ability to do this and much more.

Lesson 3: Introduction to ArcGIS – ArcCatalog

In the first two (2) lessons of this unit, you explored ArcMap as a program for displaying and analyzing geographic data. Specifically, you gained experience navigating the program in preparation for the skills, tools and techniques that you will be introduced to in the next few course units. In this lesson, you will be introduced to ArcCatalog, the program that allows you to preview geographic data prior to use in a GIS project and to manage geographic data. ArcCatalog provides a wealth of capabilities in terms of data management. This lesson will give you experience navigating ArcCatalog and learning more about its capabilities. In future lessons, you will utilize ArcCatalog to manage existing data, create new data and manage/edit data properties.

Launching ArcCatalog:
Just as you did with ArcMap, ArcCatalog can be launched from the computer desktop or from the list of programs that can be accessed from your Windows Start button.

1. *Launch* **ArcCatalog** by either *double-clicking* the shortcut on your computer desktop or by clicking the Windows Start button **start**, *pointing* your mouse to **Programs/ArcGIS** and *selecting* **ArcCatalog**. The ArcCatalog window will open.

    Take a minute to refer to the different parts of the ArcCatalog window that were introduced in the lesson presentation. Make a note of where the toolbars and other components of the ArcCatalog window are located.

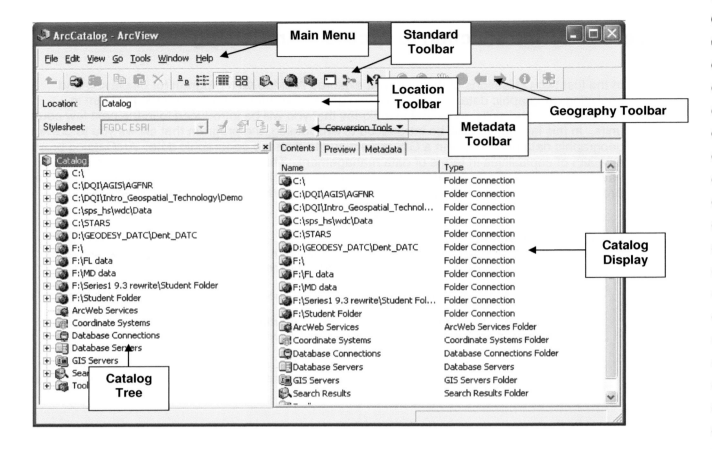

Viewing Data Contents in ArcCatalog:
In ArcCatalog, the contents of a data folder can be viewed by clicking on that folder in the **Catalog Tree** and viewing the contents in the **Catalog Display**. Data files can be seen in the **Catalog Tree** when the contents of the folder are shown. This is signified by a ⊟ to the left of the data folder in the **Catalog Tree**. Data folders that contain data files that are hidden in the **Catalog Tree** are signified by a ⊞ to the left of the data folder in the **Catalog Tree**.

2.   In the **Catalog Tree**, *navigate* to the **C:\ STARS\GIS_RS_Tools** folder. Notice the contents of the folder in the **Catalog Display**. Notice the different icons that are included to the left of the file names. Each of these icons symbolize the type of data that the data file is.

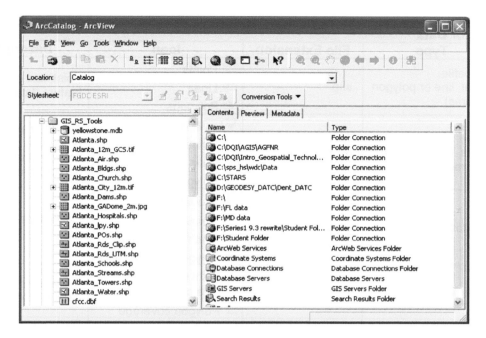

There are many different types of data files that can be used in ArcGIS programs.

3. ***Click*** the `Type` header in the **Catalog Display** to show the data files contained in this folder by their type.

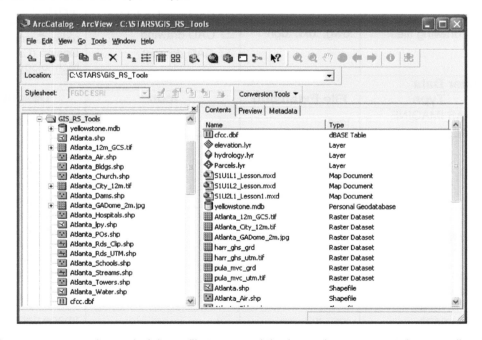

The most commonly used of these file types and the icons that represent them are listed in the following table. Keep in mind that this is not an exhaustive list, but merely a list of most commonly used data types.

Vector Data			
Type	**File Extension**	**Icon**	**Description**
Shapefile (Point, line or polygon features)	.shp	Points Lines Polygons	Point, line or polygon feature data
Layer (Point, line or polygon features)	.lyr	Layer dataset Points Lines Polygons	Files that store all the properties of the feature layer, including data source, symbology, etc.
Geodatabases	.mdb	Geodatabase Feature dataset Line feature class Polygon feature class	Collections of feature data
Coverages	None	Point coverage Line coverage Polygon coverage	Collections of feature data with topology
TIN (triangulated irregular network)	None		Data made up of nodes (points), edges (lines) and faces (polygons) to show surface data three-dimensionally
CAD drawing files	.dwg, .dxf, .dgn	Contained in folder CAD drawing CAD dataset	Drawings (usually architectural or cadastral) created using Computer Aided Design (CAD) technology
Raster Data			
Type	**File Extension**	**Icon**	**Description**
ERDAS IMAGINE image	.img		Image data layer
GRID	None		Grid surface data depicting continuous data such as elevation, precipitation, temperatures, etc.
Layer file (raster)	.lyr		Files that store all the properties of the raster layer, including data source, symbology, etc.
Bitmap image	.bmp		Image data layer
Graphic Interchange Format image	.gif		Image data layer
JPEG image	.jpg, .jp2		Image data layer
Tagged Image File Format image	.tif		Image data layer
Other Data			
Type	**File Extension**	**Icon**	**Description**

| Text files | .txt, .asc, .csv, .tab | 📄 | ASCII text-formatted data |

Other Data (continued)			
Type	**File Extension**	**Icon**	**Description**
Database files	.dbf	▦	Database table
Coordinate system	.prj	🌐	File that contains coordinate system and map projection information for the spatial data file with the same filename as the .prj file

3. On the **Standard Toolbar**, notice that the **Details** button  is depressed. The **Details** option shows the type of data layer in addition to the data layer filename.

4. ***Click*** the **Large Icons** button to see how the data contents are now displayed.

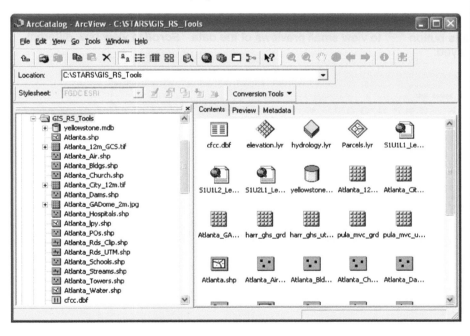

Tool Tidbit

Details ▦ - This button allows you to view the type of data layer as well as the filename of a dataset in ArcCatalog.

Large Icons - This button allows you to display the contents of a data folder as large icons in the display in ArcCatalog.

List - This button allows you to display the contents of a data folder as a list in the display in ArcCatalog.

Thumbnails ▦ - This button allows you to display the contents of a data folder as thumbnail images in the display in ArcCatalog.

5. ***Click*** the **List** button to view the folder contents as a list.

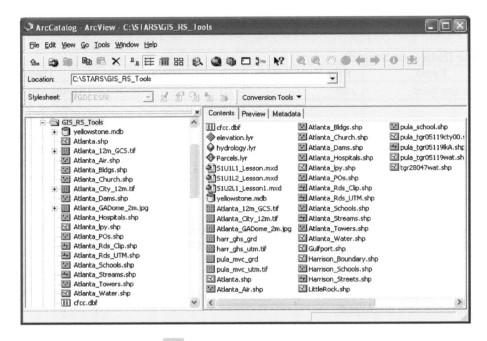

6.   ***Click*** the **Thumbnails** button [⊞] to view small previews of the data. ***Scroll*** to view how different data types are displayed.

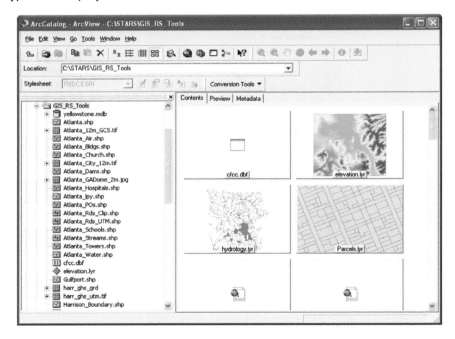

Notice that layer files actually preview the data layer for the thumbnail while shapefiles, grids and other data files show icons representing the type of data files they are.

Previewing Data in ArcCatalog:

ArcCatalog provides the convenience of allowing you to preview data layers without having to add the data to ArcMap. This saves time and allows you to better determine the data that you may need for your GIS project without having to add and remove needless amounts of data from your ArcMap map document.

7. In the **Table of Contents**, *click* on the **Atlanta.shp** data file. In the **Catalog Display**, an icon representing this data file will appear as well as the filename and the type of data file it is will be listed. Notice that this information is listed under the Contents tab in the **Catalog Display**.

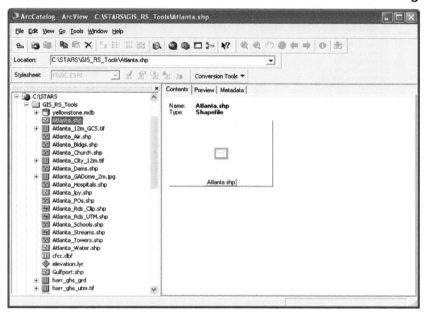

8. ***Click*** the Preview tab at the top of the **Catalog Display**. This will show you a preview of what the actual data contained in the file looks like.

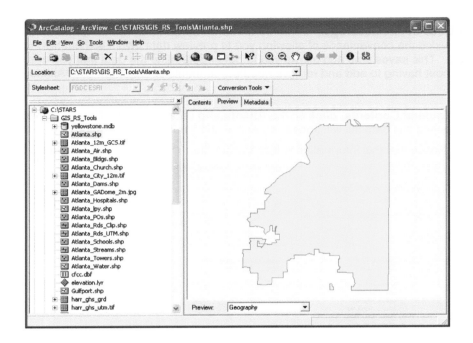

Notice at the bottom of the **Catalog Display** the **Preview option** is set as **Geography**. This means that the actual features contained in the data file will be previewed.

9. *Change* the **Preview option** at the bottom of the **Catalog Display** to **Table**. The attribute table affiliated with this data layer will display. There is only one (1) record in the table because there is only one (1) feature in this data layer – the Atlanta, GA city boundary.

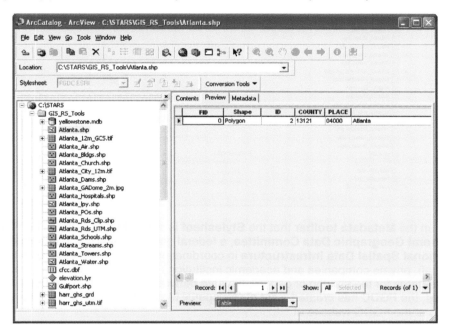

Viewing Metadata in ArcCatalog:

Metadata is information about data. Metadata provides a description about data that you may be interested in using in a GIS project. Some of this information includes the subject of the data, the type of data included, the purpose of the data, the date the data was collected, the geographic extent of the data, spatial reference (coordinate system and map projection), contact information, etc. When you are looking for data for a GIS project, you can review the metadata to determine if the data would be useful or not. If after reviewing the metadata you determine that the data would be useful, you can contact the person or organization managing the data to see about acquiring it.

10. In the **Catalog Tree**, *click* the **elevation.lyr** data file.

11. *Click* the Metadata tab at the top of the **Catalog Display**. This will show you the metadata associated with this data file.

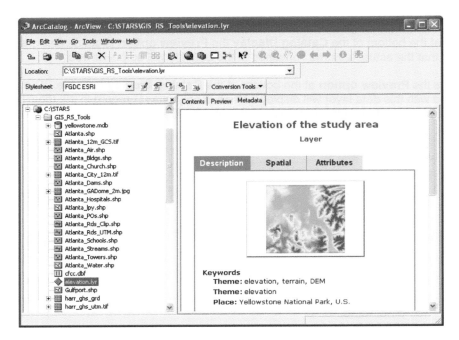

Notice on the **Metadata toolbar** that the **Stylesheet** is set as **FGDC ESRI**. The **FGDC** refers to the **Federal Geographic Data Committee**, a federal governmental committee that is developing the **National Spatial Data Infrastructure** in coordination with governmental bodies (state, local and tribal), private companies and academic institutions in order to develop policies, procedures and standards for the sharing of geographic data between organizations and entities. Through its activities, the **FGDC** has created a set of Geospatial Metadata Standards. These standards can be seen at the FGDC website at http://www.fgdc.gov/metadata/csdgm/.

In terms of these **FGDC** metadata standards, there are different stylesheets for displaying metadata in documentation. The actual data is contained in eXtensible Markup Language (XML) files and the stylesheets are used to influence how that data is presented. The FGDC recognizes **stylesheets** including **FGDC Classic**, **FGDC ESRI** (which is the default in ArcGIS because ESRI is the maker of ArcGIS software products), **FGDC FAQ**, and **FGDC Geography Network**. These are included as stylesheets in ArcCatalog.

An example of each of these types of metadata styles is included below:

FGDC ESRI Style

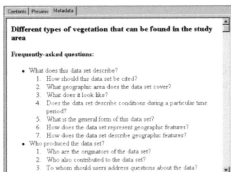

FGDC FAQ Style

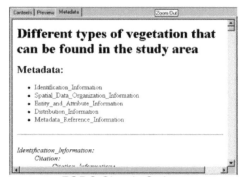

FGDC Geography Network Style   FGDC Classic Style

Stylesheets included in ArcCatalog that are preceded by the initials ISO are stylesheets that meet standards set forth by the International Standards Organization, a federation of international governments, businesses and institutions. For the purpose of our studies, we will deal strictly with FGDC stylesheets.

12. In the **Catalog Display**, look at the metadata associated with the **elevation.lyr** data file.

   Scroll down to see the **Description** information associated with this data layer. In order to view the Status of the data, Time period for which the data is relevant and other information, you may need to *click* the header for that section to view the information.

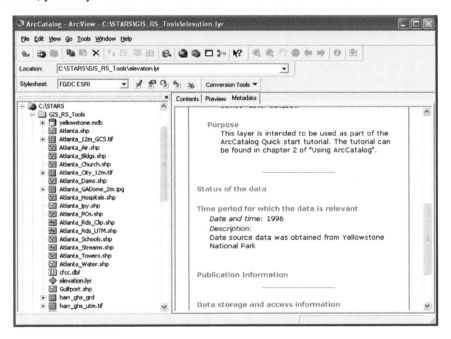

13. At the top of the metadata display, *click* the **Spatial** tab. This portion of the metadata contains information about the spatial reference of the data. Notice that the bounding coordinates – the coordinates of the north, south, east and west boundaries that form the geographic extent of the dataset – are listed in decimal degrees. In cases where data is projected, information about the coordinate system and projected coordinates is included here also.

14. On the Metadata toolbar, *change* the **Stylesheet** to **FGDC Classic**.

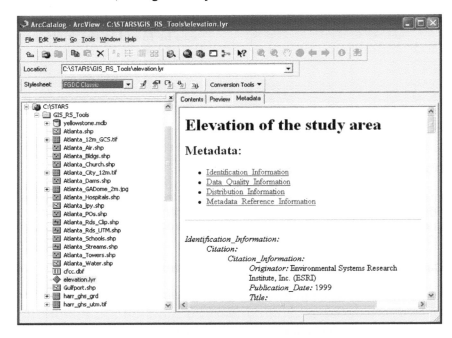

15. *Scroll* to view the metadata in this format. The same information is displayed here that you viewed earlier – the information is just displayed differently.

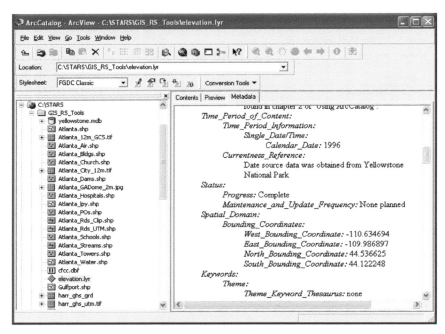

16. *Explore* some of the other **Metadata stylesheets** by changing the **Stylesheet** on the **Metadata toolbar**.

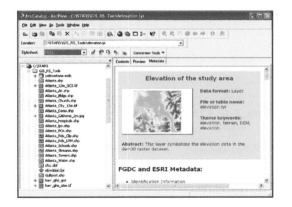 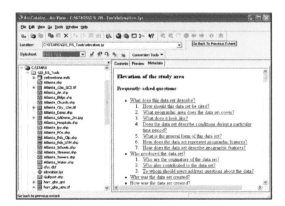

16. **Click** on the **Atlanta_Rds_UTM.shp** file in the **Catalog Tree**. **Change** the **metadata stylesheet** to **FGDC ESRI**, if necessary. **Click** on the **Spatial** tab to view spatial metadata.

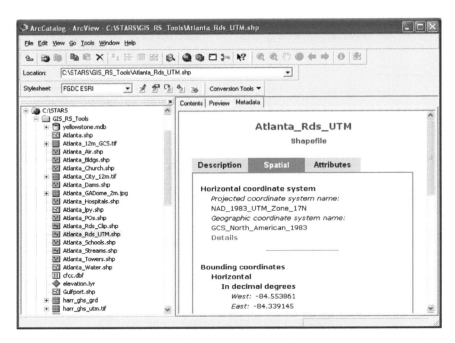

Notice that this shapefile originally utilized the Geographic Coordinate System, North American Datum 1983 (GCS_North_American_1983) but was projected using the Universal Transverse Mercator (UTM) Coordinate System Zone 17, North American Datum 1983 (NAD_1983_UTM_Zone_17N). Bounding coordinates are provided in both decimal degrees and in projected units (that for UTM would be meters).

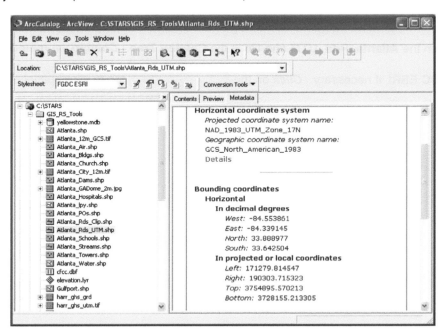

17. ***Click*** on the **Attributes** tab. ***Scroll*** to view the different attributes that are included in the attribute table for this data layer.

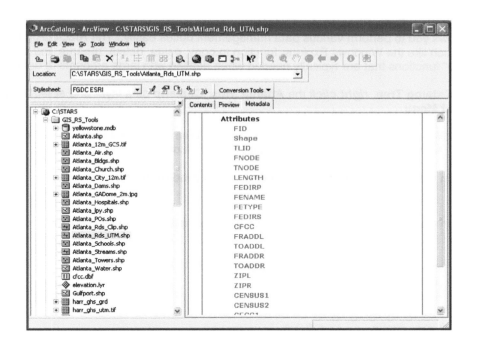

Copying & Pasting Data Files in ArcCatalog:
ArcCatalog allows you to perform various data management duties easily. The ArcCatalog **Catalog Tree** can be used much like the Windows Explorer function to copy, paste, delete, rename and perform other "house keeping" functions that are typically needed when dealing with large amounts of data.

18.   In the **Catalog Tree**, *right click* the **Atlanta_Rds_Clip.shp** data layer and *select* Copy from the **Context Menu**.

19.   *Right click* on your **student folder** in the **Catalog Tree** and *select* Paste from the **Context Menu**. A copy of the data file will be pasted in your student folder.

20.   *Click* on the **data file that you just pasted** in your **student folder**.

21.   *Click* on the **Description** tab. Notice that descriptive metadata is not included for this data layer.

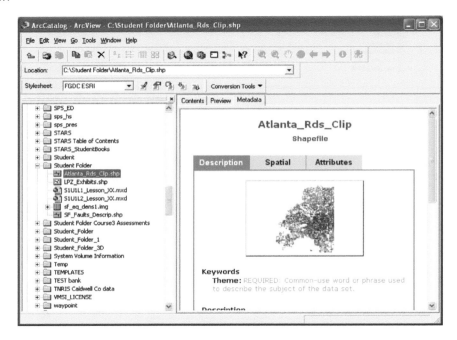

You will now add this information to the metadata for this data layer.

Editing Metadata in ArcCatalog:
In addition to viewing metadata for geospatial data files, ArcCatalog allows you to edit existing metadata. When you create new data, ArcCatalog allows you to add metadata for the new data files.

22.   On the **Metadata toolbar**, *click* the **Edit Metadata** button ![icon]. The **Editing 'Atlanta_Rds_Clip'** dialog box will appear.

Tool Tidbit

Edit Metadata ![icon] - This toolbar button allows you to change or add metadata about a specific data layer in ArcCatalog.

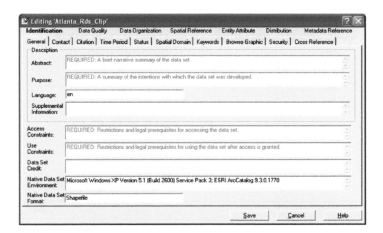

20.   *Click* in the **Abstract** box, *highlight* the text that is currently in the box and *type* the following: **This data layer contains line segments contained by the Atlanta, GA street network**.

In the **Purpose** box, *replace* the current text with the following: **This data is included as part of lesson activities created for SPACESTARS geospatial technology coursework**.

In the **Access Constraints** box, *enter* the text **None**.

In the **Use Constraints** box, *enter* the following: **Use of this data is intended for educational purposes only**.

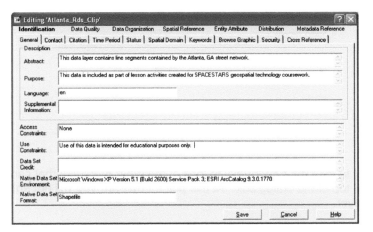

21.   *Click* the **Keywords** tab at the top of the dialog box.

51

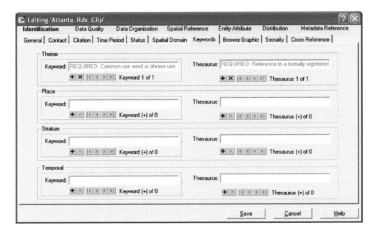

Highlight the current text in the **Keyword** box and **type** the following new text: **Streets**.

Click +. **Type** **Roads** and **click** +.

In the **Place** section, **click** in the **Keyword** box and type **Atlanta, GA**.

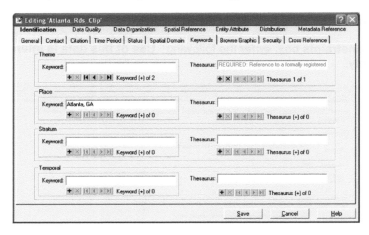

You may want to take the opportunity to look at the other metadata that can be added by clicking the tabs at the top of the dialog box. At this point, however, you will not enter any more metadata for this data layer.

When you have finished, **click** Save. The changes to the descriptive metadata will be made and displayed in the **Catalog Display**.

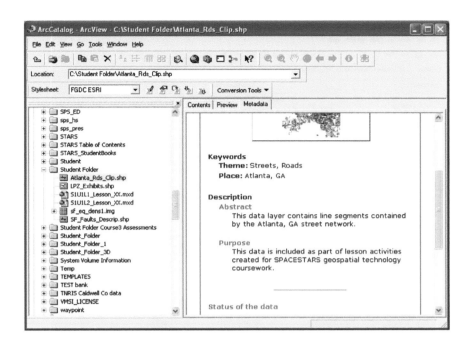

Viewing Data Attributes in ArcCatalog:

In the first lesson of this unit, you briefly viewed the attribute table of a data layer in ArcMap. ArcCatalog allows you to view attribute information as well.

22.   With the **Atlanta_Rds_Clip.shp** file still selected in the **Catalog Tree**, *click* the

Attributes tab in the **Metadata** section.

Review the attributes listed for this data layer. These attributes act as fieldnames in the attribute table.

23.   *Click* the Preview tab in the **Catalog Display**. The **preview** option at the bottom of the **Catalog Display** should be set to show the layer's **geography**. *Change* this option to **Table**. Notice the fieldnames in the attribute table match the attributes listed in the metadata section.

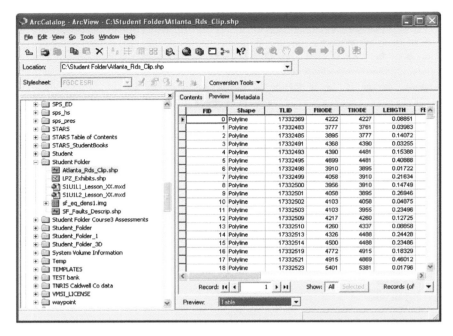

Creating a Thumbnail Image:

Earlier in this lesson, you used the **Thumbnail** display option for data files contained in a specific data folder. You may remember that most of the data layers used default data type icons instead of actual thumbnail (miniature) image of the data geography in the contents display. In this exercise, you will create a thumbnail for a data layer.

24.   ***Click*** the Contents tab.

25.   ***Click*** on your **student folder** in the **Catalog Tree** so that all of the contents of the folder are shown in the **Catalog Display**.

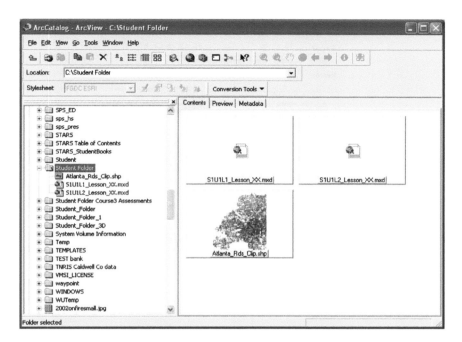

Notice in this example that the **Atlanta_Rds_Clip.shp** data layer displays as a thumbnail image. If your file does not show as a thumbnail image, it is easy to change.

26. **Click** the `Preview` tab in the **Catalog Display** and **click** on the **Atlanta_Rds_Clip.shp** data layer (in your **student folder**) in the **Catalog Tree**.

27. **Click** the **Create Thumbnail** button 🔲 on the **Geography toolbar**.

28. **Switch** back to `Contents` in the **Catalog Display** and notice that the data layer now displays a thumbnail image of the dataset instead of the default shapefile icon.

Tool Tidbit

Create Thumbnail 🔲 - This toolbar button allows you to create a miniature image of the data layer geography to display in the contents display in ArcCatalog.

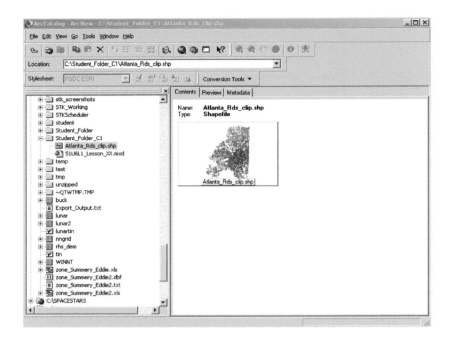

Browsing Data in ArcCatalog:

ArcCatalog contains many of the same navigational tools as ArcMap. While displaying data in Preview mode in ArcCatalog, you can zoom in and out on the data as well as identify specific features contained in the dataset.

29.   ***Click*** the Preview tab in the **Catalog Display**

30.   Use the **Zoom In** tool on the **Geography toolbar** to zoom into an area in the **Atlanta_Rds_clip.shp** data layer.

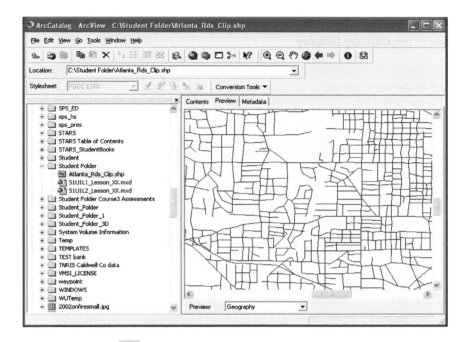

31.   ***Select*** the **Identify** tool and use your mouse to ***click*** on one of the line segments in the street network. The **Identify Results** dialog box will appear to provide attribute information about the street network segment that you chose. (Notice the fieldnames in the dialog box. They match the attributes that you reviewed in the metadata section and those fieldnames in the attribute table that you previewed.)

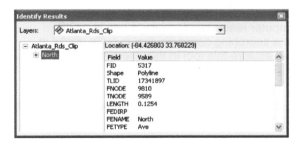

32.   ***Click*** other street segments and how the results change in the dialog box.

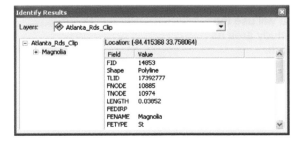

33.   When you have clicked on several line segments and explored the **Identify** tool in **ArcCatalog**, ***close*** the **Identify Results** dialog box by clicking the in the upper right hand corner of the dialog box.

---

34. ***Click*** the **Full Extent** button  on the **Geography toolbar** to zoom out to the full extent of the data layer in the **Catalog Display**.

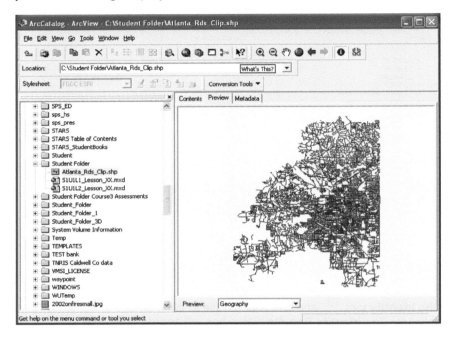

Searching for Data in ArcCatalog:

As your library of geographic data grows, there may be times when you need to search your own computer resources for data to use in a GIS project. ArcCatalog contains a searching capability that is quite useful in these circumstances. You will explore this tool in this exercise by searching for more data for the Atlanta, GA area.

Tool Tidbit

Search 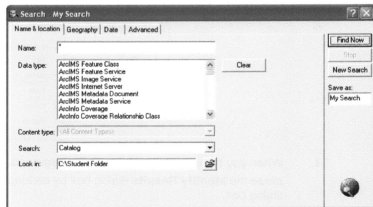 - This toolbar button allows you to search for data on your computer based on search criteria you provide.

35. ***Click*** the **Search** button on the **Standard toolbar**. The **Search – My Search** dialog box will appear.

You can search for files by filename or particular file type (as you can using the Search feature in Windows).

For this exercise, you will use the **Geography** of your study area as the basis for the search.

36. Before you set the geographic search criterion, *specify* to search the **file system** and to **look in** the **C:\STARS\GIS_RS_Tools** folder.

    You can **Browse** 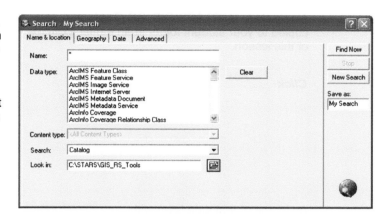 and navigate to this folder if you do not want to type the data path into the box directly.

37. *Click* the **Geography** tab in the **Search** dialog box.

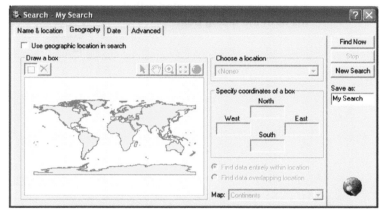

38. *Click* to

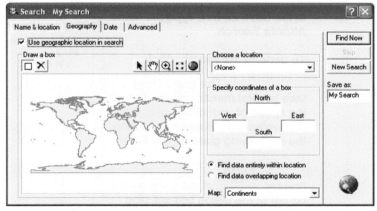

☑ Use geographic location in search . All of the options contained in this section of the **Search** dialog box should now become functional.

39. At the bottom of the dialog box, *change* the **map** option from **Continents** to **<Other...>**. This option will allow you to choose a data layer to use for the spatial extent for the search. The **Choose data to use in the search map** dialog box will appear.

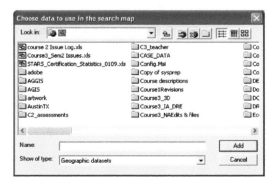

40.   ***Navigate*** to **C:\STARS\GIS_RS_Tools** and ***select*** the **Atlanta_Rds_clip.shp** file that you
copied to your student folder in an earlier exercise
of this lesson.

Click ___Add___ .

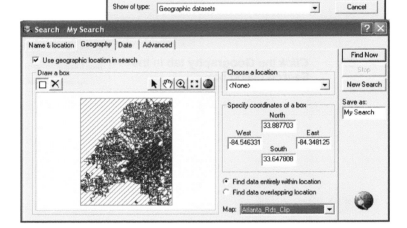

An image of that data layer with
the coordinates associated with it
will fill in the dialog box.

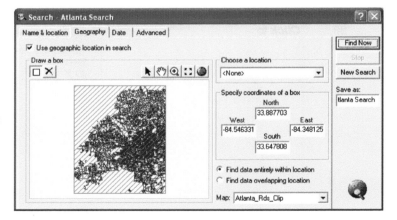

41.   In the **Save as** box, ***type*** the name
Atlanta Search.

42.   To perform the search, ***click***
Find Now . The search may
take several minutes depending on
the speed of your computer. When
the system has stopped searching,
the magnifying glass on the Earth

graphic   at the bottom right
corner of the dialog box will stop
moving.

The Search results will appear in **Search Results** in the **Catalog Tree**.

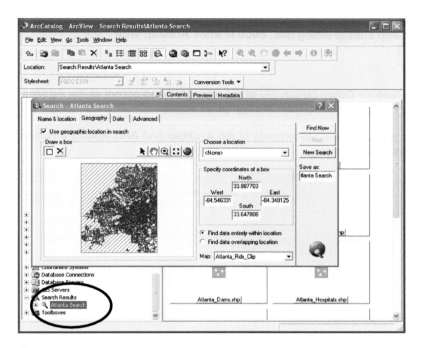

43.   ***Close*** the **Search** dialog box.

44.   ***Click*** the **Atlanta Search** in the **Catalog Tree** to display the data files that meet the geographic search criterion. At least 13 files should be listed as contents of this search. All of these files meet the search criterion of being found within the **Atlanta_Rds_UTM.shp** boundaries.

Adding Data to ArcMap from ArcCatalog:

When you have previewed data in ArcCatalog, you may decide that you want to immediately add a data layer to a new or existing map document. If this is the case, you can simply drag data layers from ArcCatalog directly into ArcMap.

45.   ***Click*** the **Launch ArcMap** button on the **Standard toolbar**. The ArcMap startup dialog box will appear.

Tool Tidbit

Launch ArcMap - This toolbar button allows you to start ArcMap from the ArcCatalog window.

47. At the **ArcMap** startup dialog box, *click* to **start ArcMap with** ⊙ A new empty map and *click* OK .

48. *Click* the **Restore Down** button [image] at the top right hand corner of the ArcMap window so that the size of the window is reduced on the computer screen.

    Adjust the sizes and the placements of the **ArcMap** and **ArcCatalog** windows so that most of both windows can be seen.

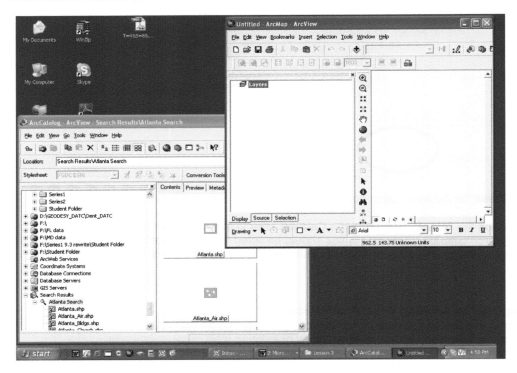

49. In the **Catalog Tree** of **ArcCatalog**, *click the* ⊞ to the left of the **Atlanta Search**, if necessary, to display the data layers in the **Catalog Tree**.)

    Click and *drag* the **Atlanta.shp** data layer from the **ArcCatalog Table of Contents** to the **ArcMap Table of Contents** OR **Map Display**. The data layer will display in ArcMap.

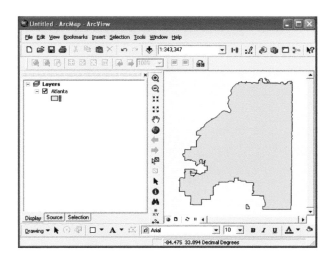

50.   *Drag* the **Atlanta_Rds_Clip.shp** data layer from ArcCatalog to ArcMap.

51.   If necessary, *drag* the **Atlanta_Rds_Clip** layer above the **Atlanta** layer in the ArcMap **Table of Contents** so that both layers can be viewed in the map display.

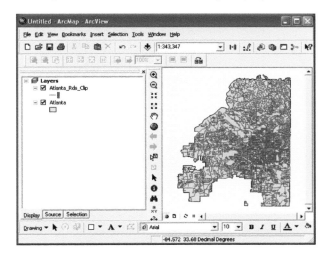

52.   In ArcMap, *save* the map document as **S1U1L3_Lesson_XX.mxd** in your **student folder**.

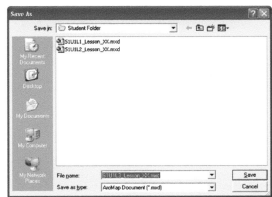

54.   *Select* **Exit** from the **ArcMap File** menu.

55.   *Exit* **ArcCatalog**.

Lesson 3: Introduction to ArcGIS – ArcArcCatalog Lesson Review

Key Terms
Use the Lesson and the Lesson Activity to define each of the following terms.

1.   ArcCatalog

2.   Metadata

3.   FDGC

4.   Catalog Tree

Global Concepts
Use the information from the lesson to answer the following questions. Use complete sentences for your answers.

5.   What are the three ways that ArcCatalog displays Metadata?

6.   Why is it easier to rename a file in ArcCatalog?

7.   What is the one thing that cannot be done in ArcCatalog?

Fill in the following two columns for each of the commonly used icons below:

Icon	Icon Name (& type)	File Extension
8. 		
9. 		
10. 		
11. 		

ArcCatalog Tools
Place the letter of the source next to the correct description. Each will be used only once.

Tool Icon		Tool Name/Function
_____12.		A. List
_____13.		B. Create Thumbnail
_____14.		C. Search
_____15.		D. Large Icons
_____16.		E. Launch ArcMap
_____17.		F. Thumbnails
_____18.		G. Details
_____19.		H. Edit Metadata

Let's Talk About It...
Answer the following question and share the responses with your instructor and classmates.

20.   Besides renaming files with ease, there are several other reasons to use ArcCatalog. What are these reasons and how do these differ from ArcMap?

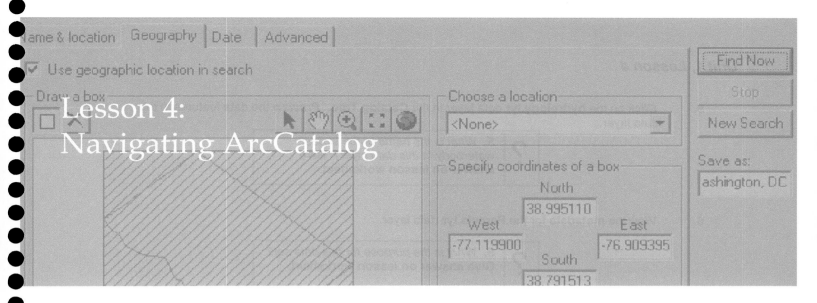

In the last lesson, you were introduced to ArcCatalog, which is one of a suite of the three ArcGIS programs you will use with this manual. ArcCatalog is the program that allows you to manage your data.

In this lesson, you will use the navigational and other skills that you learned in Lesson 3 to answer a series of questions. If necessary, refer to Lesson 3 for detailed information on ArcCatalog tools. Record your answers on the S1U1L4 Lesson Worksheet that accompanies this lesson in this manual.

1.   *Launch* **ArcCatalog** by either *double-clicking* the shortcut on your computer desktop or by clicking the Windows Start button **start**, *pointing* your mouse to **Programs/ArcGIS** and *selecting* **ArcCatalog**. The ArcCatalog window will open.

2.   In the Catalog Tree, *navigate* to the **C:\STARS\GIS_RS_Tools** folder.

> **?** *1. Name a point, a line and a polygon shapefile contained in this folder.* **Give answer on lesson worksheet**.

> **?** *2. Name two (2) raster data layers contained in this folder.* **Give answer on lesson worksheet**.

3.   *Click* to show the **Thumbnails** of the data contained in the **C:\STARS\GIS_RS_Tools** folder.

> **?** *3. What data files have an actual thumbnail associated with them?* **Give answer on lesson worksheet**.

4.   *Click* to show the **Details** of the data contained in the **C:\STARS\GIS_RS_Tools** folder.

> **?** *4. What type of data file is the Parcels layer?* **Give answer on lesson worksheet**.

Spatial Technology And Remote Sensing
STARS
Remote Sensing Education
SPACESTARS
Laboratory for GIS/Remote Sensing Education

5.   *Click* on the **hydrology.lyr** data layer in the **Catalog Tree**. *Preview* the data features in this data layer.

> **?** *5. What is the name of the largest waterbody in this data layer?* **Give answer on lesson worksheet**.

6.   *View* the **metadata** for the **Parcels.lyr** data layer.

> **?** *6. What is the purpose for this data set?* **Give answer on lesson worksheet**.

7.   *View* the **metadata** for the **harr_ghs_utm.tif** data layer.

> **?** *7. What is the projected coordinate system of this layer?* **Give answer on lesson worksheet**.

> **?** *8. What are the coordinates in decimal degrees and the projected coordinates for this data layer?* **Give answer on lesson worksheet**.

8.   *View* the **metadata** for the **Atlanta_POs.shp** data layer.

> **?** *9. List the attributes for this data layer.* **Give answer on lesson worksheet**.

9.   *Search* for data using the geography of the **washdc_cty.shp** layer (the city boundary of Washington, DC). *Save* this search as **Washington, DC**. *Select* to **Find data overlapping location**.

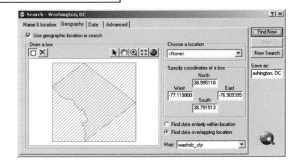

> **?** *10. How many files were found matching the criteria? What are the filenames?* **Give answer on lesson worksheet**.

10.   *Exit* **ArcCatalog**.

Lesson 4: Navigating ArcCatalog

*Name:*_____

As you are prompted in your lesson activity, answer the following questions regarding ArcMap tools and navigational procedures that you perform in the lesson exercises.

1. **Name a point, a line and a polygon shapefile contained in the C:\STARS\GIS_RS_Tools folder.**

2. **Name two (2) raster data layers contained in the C:\STARS\GIS_RS_Tools folder.**

3. **What data files in the C:\STARS\GIS_RS_Tools folder have an actual thumbnail associated with them?**

4. **What type of data file is the Parcels layer?**

5. **What is the name of the largest waterbody in the hydrology.lyr data layer?**

6. **According to its metadata, what is the purpose for the Parcels.lyr data set?**

---

7.   **What is the projected coordinate system of the harr_ghs_utm.tif layer?**

8.   **What are the coordinates in decimal degrees and the projected coordinates for the harr_ghs_utm.tif layer?**

Coordinates in decimal degrees: Projected coordinates:

West: _____   Left: _____

East: _____   Right: _____

North: _____   Top: _____

South: _____   Bottom: _____

9.   **List the attributes for the Atlanta_POs.shp data layer.**

10. **How many files were found matching the search criteria? What are the filenames?**

Managing a Data Inventory

Learning Topics

Displaying Geospatial Data

Managing Geospatial Data

Creating Geospatial Data

Analyzing Geospatial Data

Preparing Geospatial Data

Planning & Building a Local Data Inventory

Introduction to
Geographic Information Systems
Tools and Processes

Unit Two

Lesson 1:
Displaying Geospatial Data

In Unit 1, you were introduced to the ArcMap application program and became familiar with navigation options in the program. In this unit, you will begin exploring the different analysis tools and techniques that are available in ArcMap and will explore their use in the context of Homeland Security.

Features & Attributes

During the course of your geospatial studies, you will come across the terms **features** and **attributes** quite often. The term features that refers to the individual points, lines or polygons layer, represent geographic data. In a school data the the individual points or entities representing schools would be features. Likewise, the term these attributes refers to the data that describes features. In the school example, attributes associated with the school features may include the school name, the address, the type of school, the phone number, the number of students, and perhaps even the coordinates of that school. The features that make up data layers can be displayed on the map display. Attributes can be viewed by opening the data layer's attribute table.

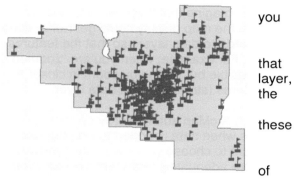

FID	Shape	DESCRIBE	CATEGORY	COUNTY	FIPS_ST	FIPS_CO	QUAD_NAME	LAT	LONG
0	Point	Abundant Life School	school	Pulaski	05	119	McAlmont	34.8389	-92.2306
1	Point	Allison School (historical)	school	Pulaski	05	119	Keo	34.5158	-92.0461
2	Point	Amboy Elementary School	school	Pulaski	05	119	North Little Rock	34.8086	-92.2922
3	Point	Argenta Elementary School	school	Pulaski	05	119	North Little Rock	34.7667	-92.2678
4	Point	Arkansas Baptist College	school	Pulaski	05	119	Little Rock	34.735	-92.2906
5	Point	Arkansas Boys State Industrial School	school	Pulaski	05	119	Woodson	34.6006	-92.1931
6	Point	Arkansas School For Deaf	school	Pulaski	05	119	North Little Rock	34.7514	-92.2992
7	Point	Arkansas School for Blind	school	Pulaski	05	119	North Little Rock	34.7525	-92.3033
8	Point	Arkansas School for Deaf and Blind	school	Pulaski	05	119	Little Rock	34.7328	-92.3264
9	Point	Arnold Drive Elementary School	school	Pulaski	05	119	Olmstead	34.8972	-92.1542
10	Point	Badgett Elementary School	school	Pulaski	05	119	Sweet Home	34.7189	-92.1925
11	Point	Baker Elementary School	school	Pulaski	05	119	Pinnacle Mountain	34.7508	-92.4383

Record: 1 of 1 Show: All Selected Records (0 out of 232 Selected) Options ▾

Each feature on a map display is associated with a record in the attribute table for that data layer. If you select a feature in the map display, the record in the table for that feature is also selected.

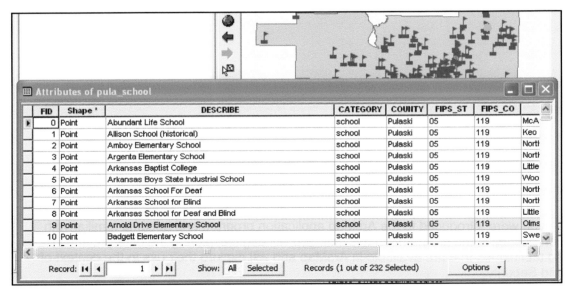

Displaying Data

You can imagine that if you were to display all of the data that you had for a certain geographic location, the map display could get rather congested. To help with this, you can apply **scale thresholds** to data layers so that the features in a layer could only be seen at certain map scales. For example, when you are "zoomed out" in a map display, the streets data layer may not be visible but when you zoom in closer than the scale threshold you would be able to view the features (streets) in that data layer.

Although ArcMap automatically symbolizes features when data is added to ArcMap, you may want to choose your own data **symbols** that more adequately represent the particular data layer. Also, when data layers are added to a map document, the filename of the data layer is used in the ArcMap Table of Contents (without the file extension). To make changes to the "symbology" or name of a data layer, you use Layer Properties. All of the options associated with how individual layers are displayed are contained in **Layer Properties**. You can access Layer Properties for the layer of interest by double-clicking the layer in the Table of Contents or by right clicking the layer in the Table of Contents and selecting Properties from the context menu that appears.

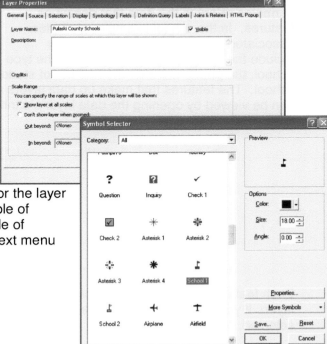

All of the data layers that are included in an ArcMap map document are organized in **data frames**. Like layer properties, data frames have certain options associated with their display as well. The default name given to data frames is Layers, but once you have accessed the **Data Frame Properties** dialog box, you have option to rename the data frame to customize it. In fact any changes that need to done to a data frame including those properties that affect ALL of the data layers in a data frame such as the coordinate system used, the display units used making measurements, the frame used to border the data frame on the map layout page, and more – can be accessed in the Data Frame Properties dialog box. You can access the data frame's properties by double-clicking the data frame in the Table of Contents or by right clicking the data frame in the Table of Contents and selecting Properties from the context menu that appears.

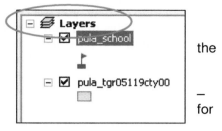

Creating the Map Layout

Once the data layers have proper symbols and names, you are ready to create a map that you can print. ArcMap allows you to create map layout pages that display geographic data to help present the findings of any analysis you have performed. There is a separate Layout toolbar that provides navigational tools specifically for the layout page. In order to add all of the "map elements" that are needed for a successful map layout – the legend, the scale bar, the north arrow, etc. – you use the ArcMap Insert menu.

ArcGIS Tools

The ArcMap Layout toolbar will appear when you switch to the layout mode. The tools contained in the layout toolbar, although they look similar to the Tools toolbar, have slightly different functions as they only work with those layers displayed in the layout window. The chart below will show you the tools you will use in this lesson compared to the similar tools found on the Tools toolbar.

Layout Tools	Use		Regular Tools	Use
Zoom In	Zooms into the layout page		Zoom In	Zooms into the data frame to more closely view data features
Zoom Out	Zooms out from the layout page		Zoom Out	Zooms out from the data features in the data frame
Pan	Allows you to pan the layout page		Pan	Allows you to pan the data features in the data frame
Fixed Zoom In	Allows you to zoom in a fixed amount to the layout page		Fixed Zoom In	Allows you to zoom in a fixed amount in the data frame
Fixed Zoom Out	Allows you to zoom out a fixed amount from the layout page		Fixed Zoom Out	Allows you to zoom out a fixed amount in the data frame
Zoom Whole Page	Allows you to view the entire layout page		Full Extent	Allows you to view the entire geographic extent in the data frame

Data Inventory

In this lesson, by combining all of the data layers to form the initial map display, you will create a data inventory for a community. This map with its data inventory is known as a **base map** and provides a glimpse of this geographic area under "normal circumstances." In terms of Homeland Security, a base map provides an accumulation of the data inventory that can be used in case of some sort of Homeland Security incident. The individual data layers can be used to perform analysis that models the event and can aid in the decision-making process. Also, the base map can provide a means for returning the community to normal after an event.

Conclusion

Features are the actual geographic entities we see on a map. For each feature you see on a map, there is an attribute table with characteristics that describe the features shown. Attributes are organized into a table in GIS projects. Layer Properties & Data Frame Properties can be used to alter how data is labeled and how it appears in a GIS project. Geospatial data and software tools can be used to create a community data inventory that can be used to create a community "base map."

This lesson will use data and resources that are pertinent to the study of Homeland Security. Homeland Security is one of the many disciplines that can benefit from the usage of the skills and techniques of GIS analysis.

Lesson 1: Displaying Geospatial Data

The key to successfully using GIS technology in Homeland Security efforts is the creation of a thorough community inventory. This inventory is made up of all of the different "resources" that are contained in the geographic area of interest whether it is your county, your city, or your neighborhood. You can think of these resources as the different types of data that describe the community – such as the locations of schools, airports, businesses, roads, streams, and recreational facilities – that is useful to us in case of an emergency. Just as a store takes inventory to know what products are in stock, a community should have an inventory as well. Stores can tell what products are missing by counting the stock on the shelf and comparing that information to the inventory list. Likewise, after a disaster, it is important to know what community features should be located there so that it is known what is "missing."

A community inventory can be used to develop a **base map** for a project. A base map simply provides the basis for a study – something to start with in the course of study. For a study in Homeland Security, a base map provides a snapshot of a community under normal, everyday circumstances.

In this lesson, you will begin to use the ArcGIS software programs to display the city of Atlanta, GA. Atlanta was chosen as the site for this demonstration project because it has many of the typical security concerns that are common for a larger urban area and also is frequently the host site of many national and international events. (You may remember some of the security issues experienced in Atlanta during the 1996 Summer Olympics.) You will create a map display of the community inventory of Atlanta, GA and create a layout of this base map.

Displaying Geographic Data using ArcMap:

1.   ***Start ArcMap*** by ***double-clicking*** the [ArcMap] shortcut on your computer desktop (or by clicking the Windows Start button [start], ***pointing*** your mouse to **Programs/ArcGIS** and ***selecting*** **ArcMap**.)

2.   When the ArcMap window appears, make sure that the radio button next to the (Start using ArcMap with) ⦿ An existing map: option is selected.

At the bottom of the ArcMap startup dialog box, make sure that `Browse for maps...` is highlighted.

Click `OK`.

Navigate to **C:\STARS\GIS_RS_Tools** and *select* the **S1U2L1_Lesson.mxd** file.

Click `Open`.

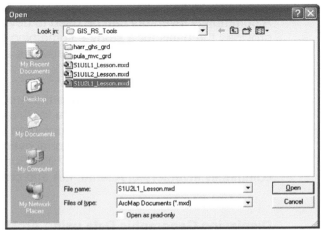

The map document will open and display data layers for Atlanta, GA.

Terminology Tip

Geographic features of the same type that are organized into sets of data that can be used in a GIS project are called *data layers*.

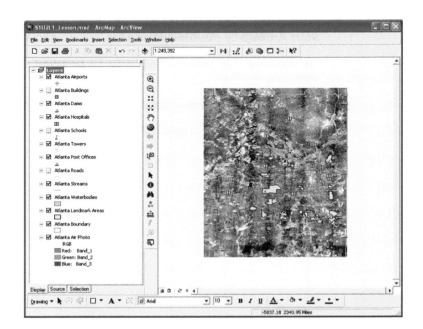

---

3. To save this ArcMap map document in your student folder, *select* **Save As...** from the **File** menu. **Name** the new file **S1U2L1_Lesson_XX** (where **XX** is your initials) and **save** the file in your student folder.

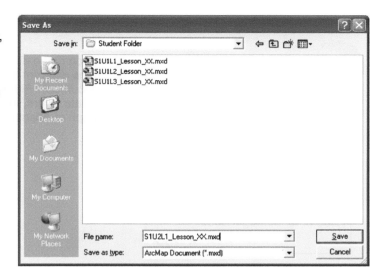

Take a minute to refer to the different parts of the ArcMap window that were documented in Unit 1 lessons. Make a note of where the toolbars and other components of the ArcMap window are located.

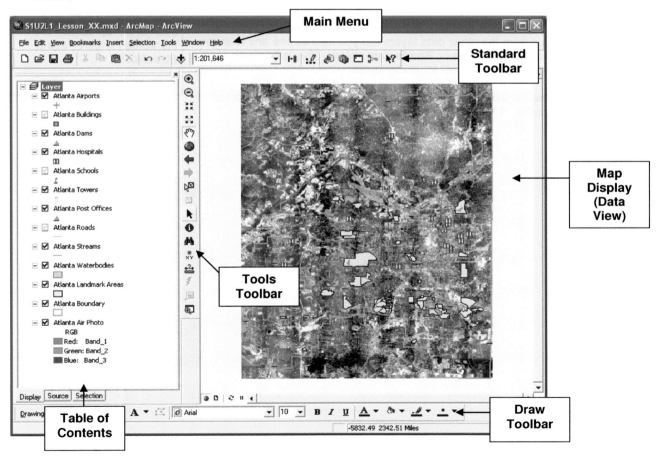

Terminology Tip

Raster data is data that is made up of equally sized cells or pixels with cell having its own pixel value. **Vector data** can be displayed as points, lines or polygons where all area within each feature has one value or is "homogeneous."

In addition to the different parts of the ArcMap window, also notice the different data layers that are included in this ArcMap map document. Although all but one of the data layers are vector data layers, there is one raster data layer, an aerial photograph of Atlanta, GA, included in this project.

Notice that all of the data layers used in this project are turned on in the Table of Contents. You can tell this because each of them has a checkmark in the box to the left of the layer name ☑. Most of the layers' checkboxes look the same with the exception of the **Schools**, **Buildings** and **Roads** layers. These layers have a grayed-out checkmark ☑. When you look at the map display, notice that you can see features from the other data layers but you cannot see any features in the **Schools**, **Buildings** or **Roads** data layers. This is because **scale ranges** have been set for these layers.

Terminology Tip

Scales settings that specify at which scales a data layer can be seen in a map display are called the *scale range* or *scale thresholds*.

4. *Use* the **Zoom In** tool to look at the area in the central portion of Atlanta. This area is identified in the following graphic.

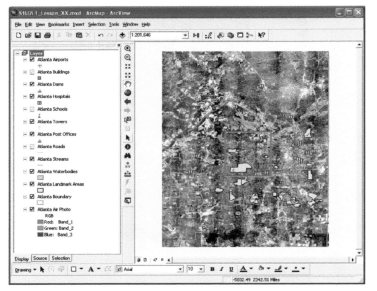

When you zoom in, you will notice that you will see features that belong in the **Schools**, **Buildings** and **Roads** layers. Also, the checkbox to the left of these data layers in the Table of Contents darkens like those of the other layers.

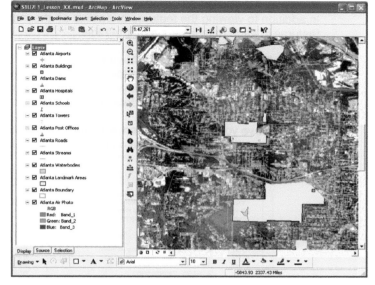

5.   ***Click*** the **Fixed Zoom Out** button to zoom out. Look at the **Manual Scale** box  as you do this. Continue to use the **Fixed Zoom Out** tool and watch the scale change and the map display change. When your scale exceeds 1:50,000, you should notice that the features in the **Schools** and **Buildings** layers no longer draw in the map display. This is because the maximum scale threshold for these layers is set at 1:50,000.

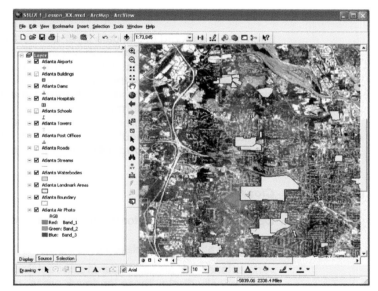

6.   ***Click*** the ***Full Extent*** tool to again see the full geographic extent of the project in the map display.

One of the most important aspects of a GIS is the fact that the geographic features that you see in a map display are not just nice symbols that have been arbitrarily placed on the map. Each of the features in a data layer corresponds to a record in a database. Each of the symbols on the map in a data layer is a **feature** or **entity**. Each feature has certain descriptive information about it known as its **attributes**.

Terminology Tip

Features are individual geographic entities represented as symbols on a map. ***Attributes*** are the descriptive qualities of features that are organized in a database table called the ***attribute table***.

Viewing Attributes:
In addition to viewing the features or entities on the map display, you can also view information about the features in each data layer's attribute table.

7.   *Right click* the **Atlanta Airports** layer in the Table of Contents. A **Context Menu** will appear. The **Context Menu** contains many common ArcMap operations. When you right click on a data layer, a **Context Menu** appears that lists common operations pertaining to data layers.

       Move your mouse to **Open Attribute Table** and *click* (left click) with your mouse to open the attribute database table for this data layer.

Each row is a record in this data layer and has a corresponding feature on the map display. Each column in the table is an attribute that describes the features in the data layer. If you scroll to the right in the table, you will see that some of the attributes include **Name**, **Type**, **County**, **County_FIP** (Federal Information Processing Standards code), **Lat_DMS** (latitude in degrees, minutes, seconds), **Long_DMS** (longitude in degrees, minutes, seconds), **Lat_DD** (latitude in decimal degrees) and **Long_DD** (longitude in decimal degrees).

Close the attribute table.

8.   *Open* the attribute table for the **Atlanta Landmark Areas**.

Notice the attributes displayed in this table. The names of the

Knowledge Knugget

Federal information processing standards codes (**FIPS** codes) are a set of codes created by the National Institute of Standards and Technology to ensure standard identification of geographic features such as states, counties, populated places, etc.

various landmark areas in Atlanta are contained in the **LANDNAME** field. To alphabetically display these landmark areas, *right click* the **LANDNAME** fieldname and *select* **Sort Ascending** from the context menu.

In the attribute table, find the record for **Piedmont Park**. *Click* the **light gray box** to the left of that row in the table so that the row is highlighted in light blue.

You have just selected a record in the **Atlanta Landmark Areas** attribute table. Now, resize and/or move the attribute table so that you can also see the map display. Notice that one of the landmark areas (Piedmont Park) on the map display is now highlighted in light blue.

This geographic feature corresponds to the record that is selected in the attribute table.

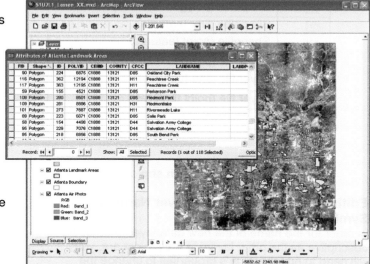

9.   *Close* the **attribute table**.

10.   To deselect the feature, click the

       Clear Selected Features
       button on the Tools toolbar.

Adding Data to a Map Display:

All of the data layers that are included in this project are included because these layers are parts of a community inventory for Atlanta, GA. You can probably imagine the importance of knowing the locations of many of these geographic features for the purpose of Homeland Security. You will now add another data layer to the project to add to the inventory base map of Atlanta. You will add a **churches** data layer.

11.   *Click* the **Add Data** button. *Navigate* to
       C:\STARS\GIS_RS_Tools.

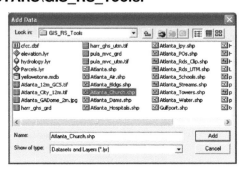

Tool Tidbit

Add Data - This toolbar button allows you to add data to an ArcMap map document.

Before selecting anything in the **Data** folder, notice the different icons that are shown in the **Add Data** dialog box. Each of these icons identify the type of dataset or layer for each file. When dealing with data for GIS projects, you may commonly see the following icons:

Icon	File Type	Icon	File Type
⧄	Shapefile (polygon features)	▦	Raster data (image or grid)
⊥	Shapefile (line features)	▤	Database table
∴	Shapefile (point features)	▤	Text table
◈	Layer file (polygon features)	▯	Personal geodatabase
◈	Layer file (line features)	⧉	Coverage (polygon)
◈	Layer file (point features)	⧉	Coverage (line)
		⧉	Coverage (point)

Select the **Atlanta_Church.shp** data layer. *Click* ___ Add ___ .

Editing Layer Properties – Layer Name & Symbology:

In order for the name of the new data layer to be better descriptive as the other layers in the project are, you will need to edit the layer's properties to change the name of the layer in the project. You can also use Layer Properties to change the symbology of the data layer. In other words, you can change the symbols used for the churches in the map display to something that better represents a church.

12. In the Table of Contents, *right click* the **Atlanta_Church** layer and *select* the **Properties** from the context menu.

If necessary, *click* the **General** tab at the top of the **Layer Properties** dialog box.

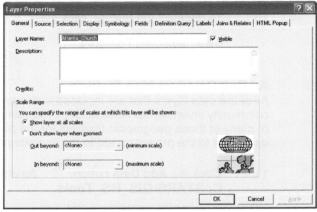

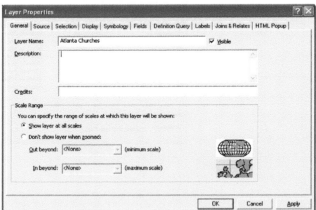

13. ***Change*** the **Layer Name** from **Atlanta_Church** to **Atlanta Churches**. (*Please note that by changing the name of the layer here, the name of the data file does not change. This simply changes to name of the layer within the context of this ArcMap map document.*)

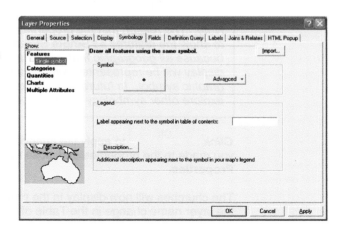

14. To change the symbols used for features in this data layer, ***click*** the **Symbology** tab at the top of the **Layer Properties** dialog box.

    Click the **symbol** that this layer was given by default _____ to open the **Symbol Selector**.

    Scroll through the symbols in the list.

    Add more symbols to the list by ***clicking*** the
    [More Symbols ▼] button and ***selecting*** **Civic**. (Additional symbols are organized by category in the [More Symbols ▼] list.)

    Again, ***scroll*** through the symbols list and now notice that additional symbols have been added.

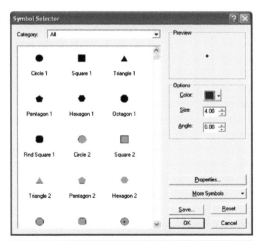

15. ***Find*** the symbol for the church [Church] in the list and ***click*** to select it.

    On the right side of **Symbol Selector** dialog box, notice that this symbol has a default color of black and size 18 point.

    Click the **down arrow** on the color button and ***change*** the color of this symbol to **light blue**. ***Change*** the **size** of the symbol to **12 point**.

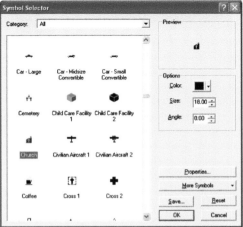

Click [OK] to close the **Symbol Selector**. The symbol you have chosen will now replace the original default point symbol in the **Layer Properties** dialog box.

> Note: The reason that the properties are set to show the features as **Single Symbol** is because all geographic features on the map display will be represented with the same graphic symbol. In future lessons you will explore other symbology show options.

Click [OK] to apply the changes that you made to the properties and close **Layer Properties**.

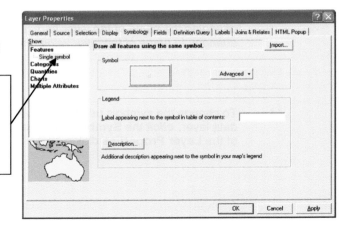

The churches will now display in the map with the changes that you made to symbology and with the layer name change in the Table of Contents.

Use the **Zoom In** Tool to see the changes to the Atlanta Churches Layer. After you are finished click the **Zoom to Full Extent** button.

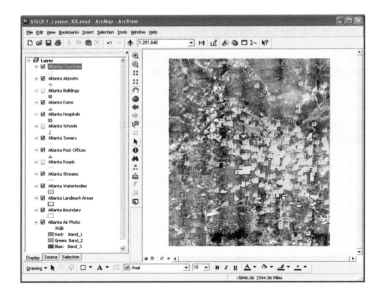

Notice that the church data layer that you just added has many features in it. In some areas of the study area, there are so many churches that the features cannot be individually distinguished. You will now apply a **scale threshold** to this layer so that this layer only appears when you zoom in to the map display.

16. ***Double click*** the **Atlanta Churches** layer in the Table of Contents to open **Layer Properties**. (You can also open Layer Properties using the context menu, as you did when you opened Layer Properties earlier in this exercise.)

   Click the **General** tab.

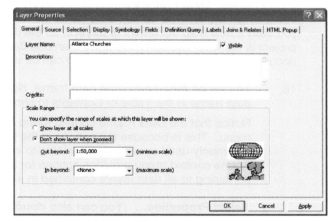

17. In the **Scale Range** section, ***click*** the radio button to specify

   ⊙ Don't show layer when zoomed: and ***enter***

   1:50,000 in the Out beyond: box.

   Click OK .

Take a moment Zoom in and Zoom Out using the various zoom tools that you have used

to see the effects when you zoom in and out between 1:50,000 scale. You can see the scale at which you are currently viewing the map on the standard

toolbar 1:151,234 . When you are finished viewing the map, click **Zoom to Full**

Extent .

How did this change the display of this data layer in the Table of Contents? What happened in the map display?

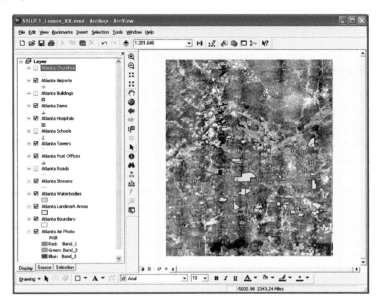

Editing Data Frame Properties – Data Frame Name:

In the last section, you opened Layer Properties for the churches data layer to edit the layer name and its symbology. As you remember from Unit 2 activities, all of the data layers that make up a single map display in ArcMap are organized into components called **Data Frames**. **Data Frames** simply group layers displayed together into a single frame. The default data frame name that appears when you start a new map document is **Layers**, as the data frame in this project is named. You can have more than one data frame in a map document; however, only one data frame can be viewed at a time.

Terminology Tip

Data layers are groups of geographic features that are of the same type. **Data frames** are ArcMap components that help you organize data by putting them in groups that are displayed together in a map display.

18. To change the name of the data frame, *right click* the **Layers** data frame in the Table of Contents to show the context menu.

Notice that the Data Frame context menu differs from the Data Layer context menu. This is because the options in the Data Layer context menu provide commonly-used operations pertaining to individual data layers and the Data Frame context menu provides options for commonly-used operations pertaining to all data layers contained in a data frame.

Select **Properties…** (You can also *double click* the **Layers** data frame to open **Data Frame Properties**.)

19. With the **General** tab in **Data Frame Properties** active, *rename* the data frame **Atlanta, GA Base Map**.

Click ⬚ OK ⬚ to apply this change and close **Data Frame Properties**.

The change is made to the data frame in the Table of Contents.

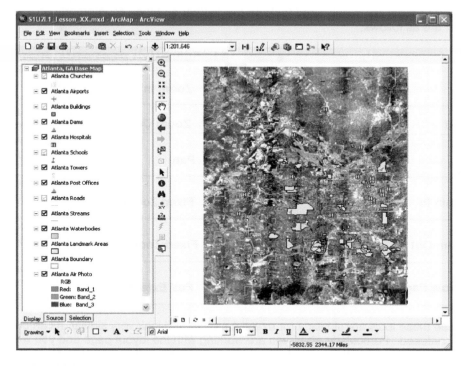

Creating a Map Layout:

In order for a map to be acceptable for publishing, it needs to be properly displayed on a map layout page and also needs to include other "map elements" – a legend, a north arrow, a scale bar, titles, etc. – that allow you to read the map properly. You will now prepare a map layout that displays the community data inventory of Atlanta, GA, otherwise known as the base map.

20.   ***Switch*** to **Layout View** by clicking on the **Layout View** button at the bottom of the Data View.

When you switch to **Layout View**, a new toolbar should automatically be added to the ArcMap Window – the **Layout toolbar**.

Terminology Tip

Data View is the mode that you use when you want to explore and query the data on the map.
Layout View is the mode that you use to prepare a map for publishing using the data frame and other necessary map elements.

Although some of the tools on the **Layout toolbar** may look similar to the tools on the regular toolbar, don't get them confused. These tools help you to navigate the **layout page**, not the data frame. The most commonly used tools on this toolbar and their counterparts on the regular toolbar are as follows:

Layout Tools	Use		Regular Tools	Use
Zoom In	Zooms into the layout page		Zoom In	Zooms into the data frame to more closely view data features
Zoom Out	Zooms out from the layout page		Zoom Out	Zooms out from the data features in the data frame
Pan	Allows you to pan the layout page		Pan	Allows you to pan the data features in the data frame
Fixed Zoom In	Allows you to zoom in a fixed amount to the layout page		Fixed Zoom In	Allows you to zoom in a fixed amount in the data frame
Fixed Zoom Out	Allows you to zoom out a fixed amount from the layout page		Fixed Zoom Out	Allows you to zoom out a fixed amount in the data frame
Zoom Whole Page	Allows you to view the entire layout page		Full Extent	Allows you to view the entire geographic extent in the data frame

Because these tools are similarly (or exactly) named and look very similar, it can be easy to get them confused. Particularly when you are working with the layout page and you want to zoom in to view the page, using the wrong tool can cause problems. If you want to zoom in or out to vary your perspective of the layout page, be sure to use the tools on the **Layout toolbar**. If you want to vary the extent of the data you view in the data frame, use the navigational tools on the regular toolbar.

21. To change the page orientation of the layout page, *select* **Page and Print Setup...** from the **File** menu.
Set Page Orientation to *Landscape*.

Place a check mark in the box next to **Use Printer Paper Settings** under Map Page Size if it is not already checked.

Click  **OK** .

22. ***Resize*** the data frame on the map layout page by clicking on the data frame box and using the sizing handles to click and drag. ***Position*** the data frame box by dragging it to a new location on the layout using your mouse.

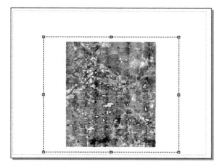

Because there are other elements that will need to be added in the map display, the frame around the map will need to be removed. After removing this frame, the image can still be resized, however, the frame border will now not be visible.

23. To remove the "frame" from the data frame on the layout page, ***double click*** the **Atlanta, GA Base Map** data frame in the Table of Contents.

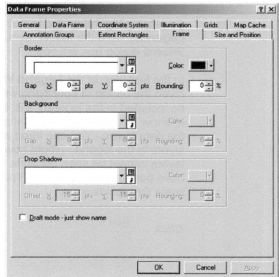

Click the **Frame** tab at the top of the **Data Frame Properties** dialog box.

Click the **down arrow** to change the border option to **<None>**.

Click OK . The single line border is removed from the data frame on the layout page.

23. To place a title on the layout page, *select* **Title** from the **Insert** menu. When the default title is placed on the layout page, enter the title **Atlanta, GA Base Map**. *Press* **Enter** on your keyboard to accept the title.

24. To change the font size of the title, *double click* the title to open **Properties**.

    Click the Change Symbol... button to open the **Symbol Selector**.

    Change the font **size** to **28** with **bold** **B** style. *Click* OK to close the **Symbol Selector**. *Click* OK in **Properties** to apply the change to the title.

25. To place a legend on the layout page, *select* **Legend...** from the **Insert** menu. The **Legend Wizard** dialog box will open.

    In the first step of the **Legend Wizard** you are given the option of specifying what data layers will be included in the legend. It automatically defaults to show all of the data layers that are included in the data frame. You can keep all of the data layers in the legend. (Right now, three of the data layers cannot be seen because scale thresholds have been applied to them. You will remove the scale thresholds soon.)

    Click Next >.

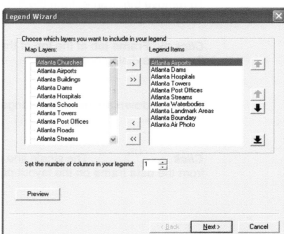

    In the next step of the **Legend Wizard**, you are prompted to choose formatting style for the legend. This includes the title of the legend and font size. Accept the defaults.

    Click Next >.

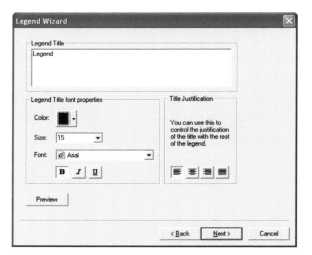

The next step of the **Legend Wizard** allows you to specify a border for the legend, if one is wanted.

Click Next > to leave the legend with no border.

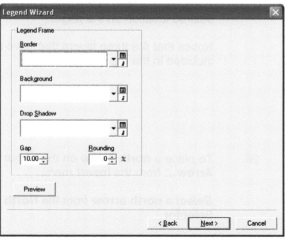

If you would like to change the size of the symbol patch that is used in the legend to show line and polygon features, you may do so in this step of the **Legend Editor**. Accept the defaults that are given.

Click Next >.

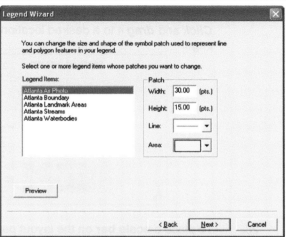

In the last step of the **Legend Editor**, you can specify spacing preferences between the different parts of the legend. Accept the defaults.

Click Finish.

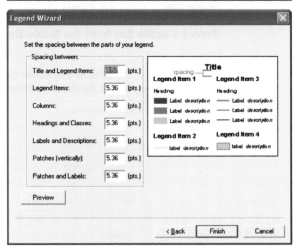

A legend is added to the layout page. *Click* and *drag* it to the desired location on the page.

Notice that the three layers that have scale thresholds are not included in the legend.

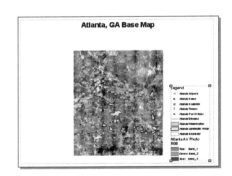

26.   To place a **north arrow** on the layout page, *select* **North Arrow...** from the **Insert** menu.

Select a **north arrow** from the **North Arrow Selector**.

Click [OK] to add the arrow to the page. *Click* and *drag* it to a desired location on the page.

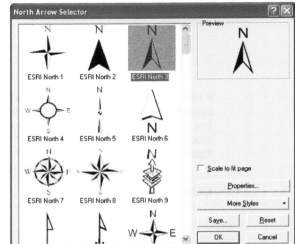

26.   To place a **scale bar** on the layout page, *select* **Scale Bar...** from the **Insert** menu.

Select a **scale bar** from the **Scale Bar Selector**.

Click [OK] to add the scale bar to the page. *Click* and *drag* it to a desired location on the page.

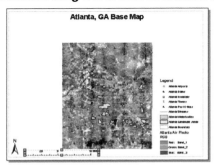

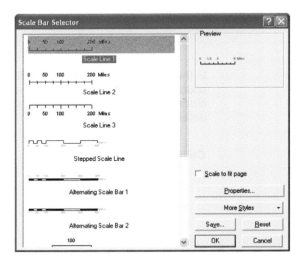

27.   To place your **name** and the **date** on the layout page, *select* **Text** from the **Insert** menu. A text box will appear on the layout page (just as the title box did). It may be difficult to see the text if it is placed on another map element on the page.

Type your **name** and *press* **Enter** on your keyboard.

Click and *drag* the text box to a desired location on the map layout page.

Double click the **text box** to open **Properties** to add the date to the text box.

Click in the box next to your name and *press* **Enter** on your keyboard to go to the next line in the text box. *Type* the **date**. When you have finished, *click* OK.

You will now add a page border to the layout page. Make sure that no elements are selected on the layout page before you proceed.

28. To add a border to the layout page, *select* **Neatline…** from the **Insert** menu. The **Neatline** dialog box will appear.

Confirm that the  option is selected. *Choose* a **border** style from the drop-down list and *select* **Hollow** as the **background**.

Resize the Neatline as necessary by placing the cursor over one of the blue boxes. Click and hold when the arrow appears and drag to the position each corner as needed.

You may also choose to place the Neatline inside margins, which will place the neatline as a boundary on the map layout.

Click OK.

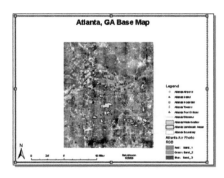

Editing a Layer after Layout:
Sometimes you may need to change some properties of a data layer after you have started page layout. This is no problem. You will now remove the scale thresholds from three layers in this project.

29. **Open** Layer Properties for the **Atlanta Churches** layer. (You can open **Layer Properties** by **double clicking** the layer in the Table of Contents or by **selecting** Properties from the context menu.)

If necessary, **click** the **General** tab. **Delete** the **50,000** from the **minimum scale** box.

Click OK .

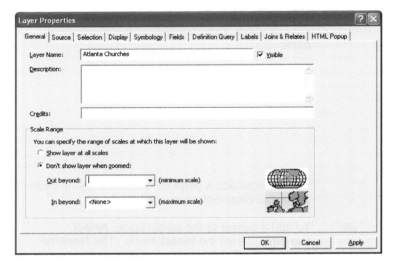

Notice that the **Washington, DC Churches** layer is now added to the legend on the layout page. However, it is added to the bottom of the legend and not at the top of the legend with the other point data layers. This is because it was the last layer added to the map document. You will fix this later.

30. **Repeat** this process to remove the **minimum scales** from the **Atlanta Schools** and **Atlanta Buildings** layers.

31. **Click**  to save the map document.

Exporting a Map Layout:
A map layout can be exported from ArcMap as an image file. By doing this, a map layout can be inserted into a word processing program or into a presentation program. This is very important when you need to create a written report or give an oral report to communicate project findings.

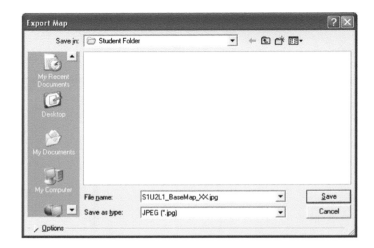

32. To export the map as an image file, **select** **Export Map...**from the **File** menu.

33. **Export** the layout to **your student folder** in **JPEG** format as **S1U2L1_BaseMap_XX** (where XX is your initials).

Click Save .

Printing a Map Layout:

34.　***Select Print...*** from the ***File*** menu to send map to the printer.

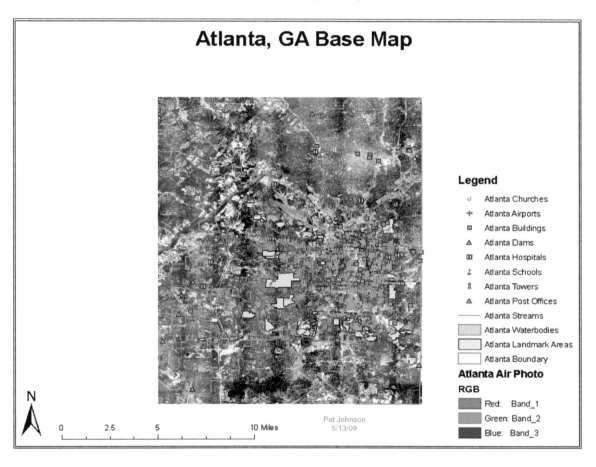

Saving & Closing:

35.　***Save*** this project.

36.　***Select*** **Exit** from the **File** menu or ***click*** the **Close** ☒ button in the upper right corner of the ArcMap window to exit.

Lesson 1: Displaying Geospatial Data Lesson Review

Key Terms

Use the Lesson and the Lesson Activity to define each of the following terms.

1. Features

2. Attributes

3. Scale Thresholds

4. Symbol

5. Feature

6. Data Layers

7. Layer Properties

8. Data Frame

9. Data Frame Properties

10. Base Map

Global Concepts

Use the information from the lesson to answer the following questions. Use complete sentences for your answers.

11. What happens to the map display when you highlight one record within an attribute table?

12. What is one benefit of setting a scale threshold?

13. Name six map elements that are added to the map layout in the Lesson Activity.

14.   What is the benefit of exporting a map as an image file?

Let's Talk About It...
Answer the following question and share the responses with your instructor and classmates.

15.   In the Lesson Activity, the Atlanta churches in the Atlanta Churches data layer were all given the same symbol with the same color. In future lessons, you will learn how to display them with different sizes and colors. How might you apply different colors and sizes to this layer?

FID	Shape*	ID	POLYID	CENID	COUNTY	CFCC	LANDNAME	LANDPOLY
0	Polygon	64	1291	C1888	13121	H41	City of East Point Reservoir	47
1	Polygon	67	1358	C1888	13121	H31	Tatum Lake	50
2	Polygon	68	1372	C1888	13121	H31	Wildwood Lake	51
3	Polygon	82	1952	C1888	13121	D82	Hollywood Cemetery	55
4	Polygon	83	1963	C1888	13121	D82	Hollywood Cemetery	55
5	Polygon	84	1974	C1888	13121	D82	Hollywood Cemetery	55
6	Polygon	85	2019	C1888	13121	D85	West Manor Park	56
7	Polygon	86	2022	C1888	13121	D85	Cascade Springs Nature Prsv	57
8	Polygon	87	2023	C1888	13121	D85	Cascade Springs Nature Prsv	57
9	Polygon	88	2045	C1888	13121	D82	Southview Cemetery	58
10	Polygon	89	2222	C1888	13121	H31	Niskey Lake	59
11	Polygon	91	2251	C1888	13121	D44	Mt Gilead Campground	61

Record: |◄ ◄ | 1 | ► ►| Show: All Selected Records (0 out of 118 Selected.) Options ▼

Lesson 2:
Managing Geospatial Data

Once a map, like a base map, is created, many different types of analyses can be performed. There will be times when the data layer files come with more information than necessary for your particular study and there might also be times when the data files do not have enough information necessary for your study. Regardless of the situation, there are methods of managing the data you have acquired to make it work for you.

Managing data
As a project manager it is possible that you may be conducting analysis on data in a way that has not ever been performed before. In cases like this, data may have to be added or tables may need to be edited to cater the data to the specific requirements of your study. Editing tables in ArcMap is not a difficult task if you have a small amount of data to edit. Let us say that you had a restaurant data file that gives the locations for restaurants but not their names. You would like to display the names of the restaurants in a map. For a very small town, for instance, you may only have 20 or fewer restaurant names to key into your restaurant attribute table. If you were working with a restaurant file for a large metropolitan area, there may be thousands of restaurant names to type in. It would be easier to find a restaurant file that included the names and join it to the file you currently had. When joining tables, you must find a field that is common to both tables. In the example, each table has a field titled CFCC (Census Feature Class Codes). These tables can be joined using this field.

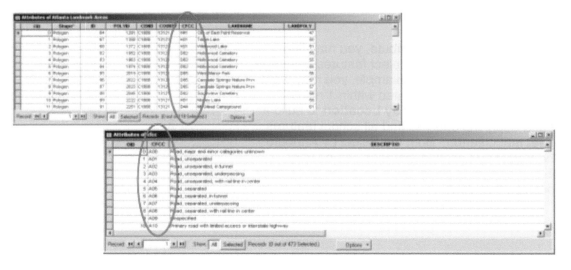

Spatial Technology And Remote Sensing
SPACE STARS
Laboratory for GIS/Remote Sensing Education

It is important to note that a spatial data layer (containing data that can be displayed on a map) can be joined to another spatial table or it can be joined to a non-spatial table (containing data that cannot be displayed on a map) using this method. In the previous example given, the table on the top is the attribute table for a Landmark Area data layer. The table at the bottom is simply a table listing CFCC codes and their descriptions. Joining these two tables would give the user the ability to analyze the different types of landmarks as designated by the US Census Bureau. Another example may be that you have two complimentary sets of data for a large area that need to be joined in order to study the entire area. Maricopa County, Arizona is one of the largest counties in the United States and it contains many, many data sets. In order to perform analysis on data for the entire county, you may have to join several files together.

Selecting features

One important capability in GIS analysis is the ability to select features in a map display. You already have some experience interactively selecting features in a map using the Select Features tool. In performing various analyses, you may have the need to select features that meet certain a certain criterion or criteria. For example, you may need to determine the locations of all interstates in a street data layer. You wouldn't want to have to interactively find each individual street segment that meets that criterion. Instead, you can use **Select by Attributes** to find these features. By entering an equation that reflects the criterion of interest, you can immediately select any features that meet that criterion. You build equations using the fields contained in the attribute table for a data layer and the "value" that satisfies your criterion. For example, if you wanted to find all highways in the street data layer and that information is contained in the feature type (called FETYPE) field of that layer's attribute table, you could build the equation FETYPE = 'Hwy,' apply that query and the features that meet that criterion would be selected in the map display (and in the attribute table).

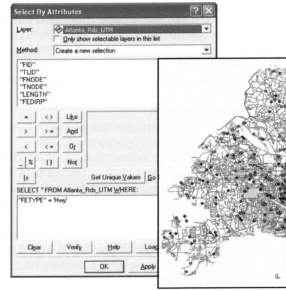

In addition to Selection by Attribute, you may find occasions when you want to select features based on where they are located. For example, you may want to find all schools that are located within 1 mile of airports. You can use **Select by Location** to build a query and apply the query to find all school features that meet that criterion. Both the Select by Attribute and the Select by Location methods can be very valuable in analyzing geographic data and geographic relationships that exist.

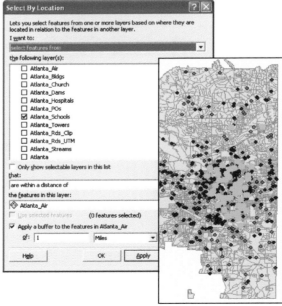

Labels

ArcMap can be used to automatically label features contained within a data layer so that they can be seen in a map display. The labels can be edited to have larger font sizes and colors and if needed, the locations of the labels can be changed as well. If you want to label all features for a data layer, right click on the layer in the Table of Contents and select Label Features. The labeling features can be also be accessed by going through Layer Properties for a layer. This method allows more editorial options such as font sizes, colors, scale range, and more.

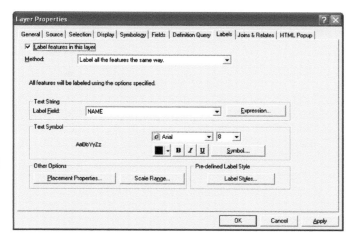

Homeland Security Scenario

In this lesson activity, you will explore selection methods using attribute-based and location-based queries involving a Homeland Security scenario. Although these tools, techniques and skills are being introduced using Homeland Security scenarios, they can also easily be used for geospatial analysis in other disciplines as well.

Conclusion

There are an infinite number of analyses that can be performed in ArcMap. The analysis in this lesson will focus on the task of Homeland Security. One technique used in this lesson involves editing tables by joining them together. This allows users to have a greater set of attributes in which to conduct their analysis. Joining tables together can only be performed with two tables that have a mutual field. Other techniques used in this lesson are the Select by Attributes and Select by Location methods. Both of these query tools provide a quick way to filter through the attributes to find specific data based on certain criteria whether it is based on a trait or a locale. All of the skills and techniques used in this lesson can be applicable to other scenarios as well.

Lesson 2: Managing Geospatial Data

Once you have developed a community data inventory, you can begin to manage that data based on your specific needs. You have probably already recognized that a community inventory can be very valuable for many applications – not just Homeland Security. From an inventory, you can simulate different scenarios and determine the best courses of action to take for situations ranging in discipline.

In this lesson and future lessons in this unit, you will explore some scenarios that are common to the topic of Homeland Security. You will be introduced to and gain experience working with the tools and techniques necessary to explore these scenarios. Keep in mind, however, that these tools and techniques are not exclusive to the study of Homeland Security.

Opening an Existing Map Document:
You will use the map document you saved from Lesson 1 for the activities contained in this lesson.

1.   ***Start* ArcMap** by ***double-clicking*** the **ArcMap** shortcut on your computer desktop (or by clicking

the Windows Start button , ***pointing*** your mouse to **Programs/ArcGIS** and ***selecting* ArcMap**.)

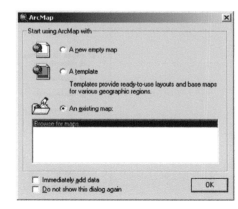

2.   When the **ArcMap** window appears, make sure that the radio

button next to the (Start using ArcMap with) **An existing map:** option is selected.

At the bottom of the ArcMap startup dialog box, make sure that **Browse for maps...** is highlighted.

Click **OK** .

---

3. ***Navigate*** to **your student folder** and *select* the **S1U2L1_Lesson_XX.mxd** file that you saved from the last lesson.

   Click Open .

The map document will open and will still be in **Layout View** from the last lesson.

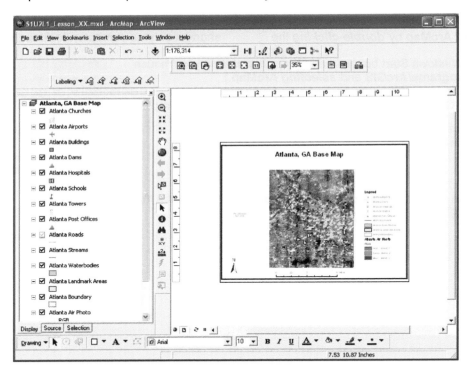

4. ***Click*** to switch back to **Data View**.

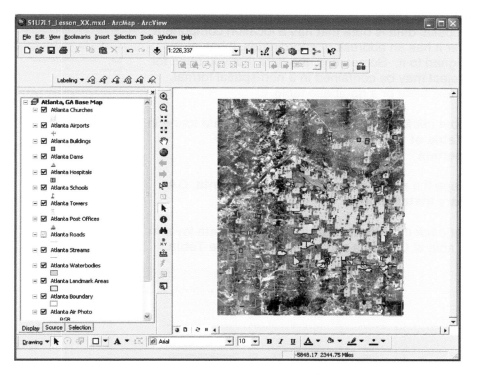

5.   ***Select* Save As...** from the **File** menu and ***save*** this
map document as **S1U2L2_Lesson_XX.mxd**
(where **XX** is your initials) in your student folder.

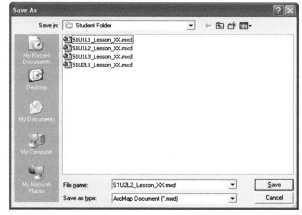

Homeland Security Scenario

The Department of Homeland Security has issued a warning for military
installations on U.S. soil. Intelligence officials have been monitoring Internet
postings by known terrorist groups. Your organization has been alerted to the
warning. You need to determine what, if any, of these installations exist in your
community in order to distribute necessary information to officials.

Contained in the community data inventory that you used in the previous lesson is a data layer
containing Atlanta landmark areas. You will use this data layer to determine if any such
installations exist in Atlanta.

Using Non-Spatial Data:
Simply put, spatial data is data that has a location component to it. Whether it has an address or coordinates, it has a location assigned to it. Sometimes, the data that you need for a specific project may be contained in a non-spatial format, such as in a text table or a database table.

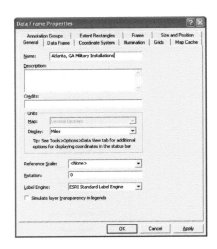

6.   *Double click* the **Atlanta, GA Base Map** data frame in the **Table of Contents** to open **Data Frame Properties**.

      Change the **name** of the data frame to **Atlanta, GA Military Installations**.

7.   *Right click* the **Atlanta Landmark Areas** data layer in the Table of Contents and *open* its **Attribute Table**.

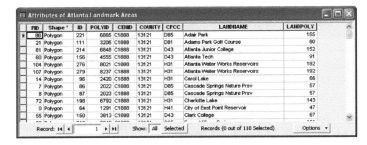

Notice that this data layer has a **Shape** fieldname in its attribute table. This field indicates that this is a spatial dataset – that each feature in the dataset has its own place within the geographic extent of study.

While you have the attribute table open, also notice that one of the attributes that is included in the table is **CFCC**. **CFCC** stands for Census Feature Class Code. The U.S. Bureau of the Census has created a list of codes that provide information on the classification of a geographic feature. A table of CFCC codes and their descriptions is available in tabular format.

By looking at the CFCC attributes in this table, it is impossible to determine what type of landmark each of the features is. In order to know that, you would need to add a description attribute to this table or look at a table that contained all of the CFCC codes. You will do this now.

Knowledge Knugget

If you open the attribute tables of other data layers such as the Roads and Waterbodies, you will see that these layers also have a CFCC attribute associated with the their features. Any data that is derived from the U.S. Census Bureau will contain CFCC code attributes.

Close the **Attributes of Atlanta Landmark Areas** table.

8.   ***Click*** the **Add Data** button  . ***Navigate*** to **C:\STARS\GIS_RS_Tools**. Notice the
      cfcc.dbf data file. This particular file is a database table of CFCC codes and their
      descriptions. ***Select*** this data file and ***click***
      Add .

When you add this table to ArcMap, you may notice
something different about the Table of Contents.
When you add a non-spatial data file to a map
document (like a text table or database table like the
cfcc.dbf file that you just added), the Table of Contents
switches to **Source mode**. Under normal circumstances,
the Table of Contents is viewed in **Display mode**,
meaning that all layers shown in the map display are
listed in the Table of Contents in the order that they are
layered in the map display. Because a non-spatial table
cannot be displayed on a map, it would not be listed in
the Table of Contents in **Display mode**. **Source mode**
shows <u>all data</u> used in the ArcMap map document by the data path where the data file is
located.

> **Terminology Tip**
>
> The *Display* tab shows the data
> layers in the Table of Contents in
> the order they are drawn in the
> map display. The *Source* tab
> shows the data in the Table of
> Contents based on the location of
> the data files.

9.   ***Right click*** the **cfcc.dbf** table in the Table of Contents and ***select*** **Open**.

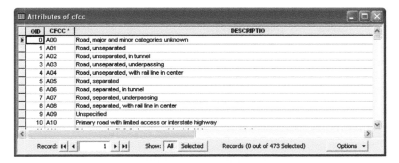

This table contains CFCC codes and the corresponding description of each. This table is
a non-spatial table. However, because this table contains CFCC codes and the **Atlanta
Landmark Areas** attribute table contains a CFCC field, you can join the two tables so that
the **Atlanta Landmark Areas** attribute table contains the descriptions of the CFCC codes
that apply to the Atlanta landmarks.

Close the **cfcc.dbf** table.

10.   ***Click*** the Display tab at the bottom of the Table of Contents to switch to **Display mode**.

11. ***Right click*** the **Atlanta Landmark Areas** data layer in the Table of Contents and ***select Joins and Relates/Join…*** The **Join Data** dialog box will appear.

    Select the **CFCC** field as the layer on which the join will be based.

    Select the **cfcc** table as the table to join to the layer.

    Select **CFCC** as the field in the table to use for the join.

    ***Select* Keep only matching records** from the Join Options section.

    Click 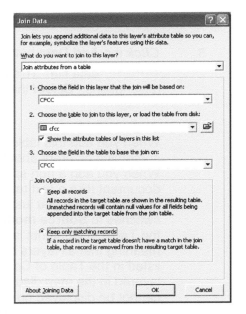 OK to perform the join.

12. ***Open*** the **attribute table** for the **Atlanta Landmark Areas** data layer. Scroll right to look at the extra fields that have been added including the **Description** field. *Note that the field name only allows for 10 characters. This is why the Description field is labeled DESCRIPTIO instead of DESCRIPTION.*

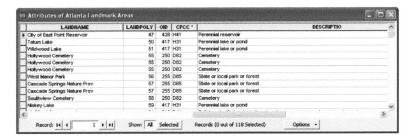

    You can now use this description data in your analysis.

13. ***Close*** the attribute table.

Selecting Features by Attribute:
Now that you have joined the CFCC description data to the landmark data layer, you will use this information to find military installations in Atlanta. You will use the **Select by Attribute** method to find these landmarks.

14. ***Choose* Select by Attributes** from the **Selection** menu.

15. In the **Select by Attributes** query box, ***select* Atlanta Landmark Areas** as the **layer**.

16. Use the **fieldnames** and the **equation operators** to ***build*** an

Knowledge Knugget

Select by Attribute is the technique used to select features in a data layer using criteria to query the features' attributes. These attributes are typically not seen on a map but are contained in the data layer's attribute table.

equation to query the database to find landmark areas that are classified by CFCC code as military installations:

Double click the **cfcc.DESCRIPTIO** field in the list. ***Click*** `=`.

Click the `Get Unique Values` button so that the list of all entries in the **DESCRIBE** field are listed.

Scroll to find **Military installation or reservation** in the list of unique values. ***Double click*** it to add it to the query equation.

Click `Apply`. Those areas will be selected in the map display. ***Click*** `OK` to close the **Select by Attributes** query box.

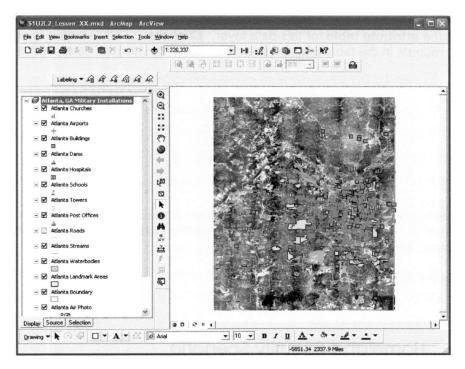

17. ***Select*** **Zoom to Selected Features** from the **Selection** menu.

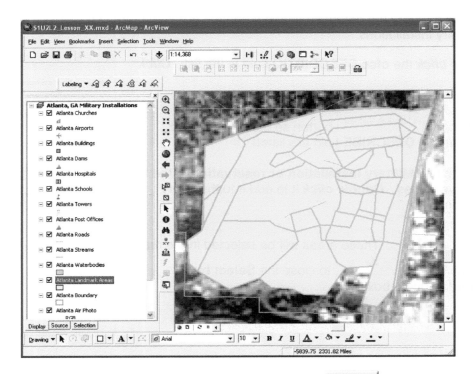

18.   ***Open*** the **attribute table** for this data layer. ***Click*** the Selected button to view only those records in the table that are selected. ***Scroll*** to view the name of the landmarks.

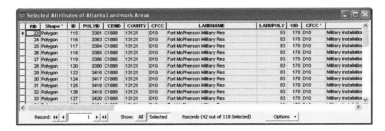

The landmark areas that are selected represent areas in the Fort McPherson Military Reservation. You will need to notify the necessary personnel of the Department of Homeland Security warning.

19.   ***Close*** the table.

Homeland Security Scenario

The Department of Homeland Security has upgraded its initial warning to military installations. From information received from DHS, you determine that locations near the risk area with potential high populations should be identified in case of emergency. You determine that any schools located within 1 mile of the site should be identified.

Selecting Features by Location:

Now that you have identified the risk area in the community, you can use the **Select by Location** option to find geographic features that are in close proximity to the risk area.

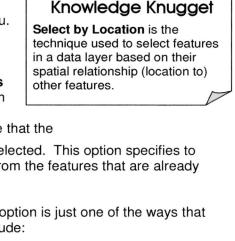

20.   *Select* **Select by Location…** from the **Selection** menu.

Knowledge Knugget

Select by Location is the technique used to select features in a data layer based on their spatial relationship (location to) other features.

Specify that you want to **select features from** the **Atlanta Schools** layer that **are within a distance** of **1 mile** from the features in the **Atlanta Landmark Areas** data layer. The **Are within a distance** selection option allows you to select features that are located within a certain specified distance of other features. Make sure that the

☑ Use selected features   (42 features selected)   option is selected. This option specifies to only find features that are located within the distance from the features that are already highlighted (or selected) in a data layer.

Keep in mind that the **Are within a distance of** query option is just one of the ways that you can select features by location. Other options include:

Intersect: Selects features that are overlapped by features of another layer and features that border the features of the other layer
Completely Contain: Selects polygons in a layer that completely contain features from another layer
Are Completely Within: Selects features in a layer that are completely inside the polygons of another layer
Have Their Centroid in: Selects polygon features in one layer that have their center in polygon features of another layer
Share a Line Segment with: Selects line and polygon features that share line segments with other features
Touch the Boundary of: Selects lines and polygons that share line segments, vertices, or end-points with the lines in the layer (but features are not selected if they cross the lines in the layer)
Are Identical to: Selects any feature that has the same geometry as a feature of another layer (feature types must be the same – point, line or polygon)
Are Crossed by the Outline of: Selects features that are overlapped by features of another layer
Contain: Selects features in one layer that contain the features of another (and boundaries of the features ARE allowed to touch)
Are Contained by: Selects features in one layer that are contained by features in another layer

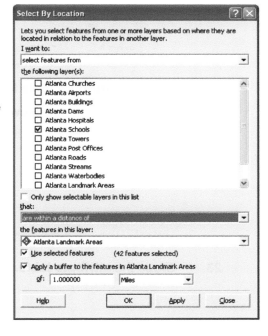

Click Apply and then *Click* OK to close the **Select by Location** query box.

21.   ***Choose* Zoom to Selected Features** from the **Selection** menu.

The schools that are located within 1 mile of the affected risk area will be selected in the map display.

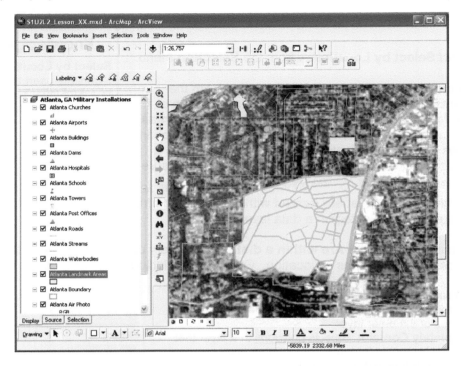

22.   ***Open* the attribute table** for the **Schools** data layer to see the selected records. ***Click*** the [Selected] button to view only the selected records.

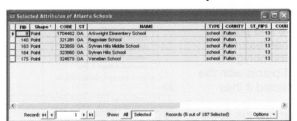

Five schools were selected that meet the one mile criterion.

23.   ***Close* the attribute table**.

Labeling Features Interactively:
You can now label the school features that have been selected so that you can tell what the school names are on the map. Although ArcMap can automatically label all features in a data layer for you (which you will do in a later lesson), only the selected features in this data layer need to be labeled. In this case, you will label these few features interactively.

24.　　First, *open* **Layer Properties** for the **Atlanta Schools** data layer by *double clicking* the layer in the **Table of Contents**.

Click the **Labels** tab. Make sure that the **NAME** field is set as the **Label Field**.

Click ⎣　OK　⎦.

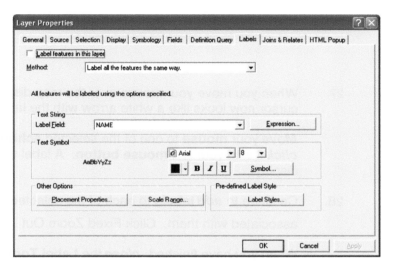

25.　　*Click* the **down arrow** ▾ to the right of the **New Text** tool on the **Drawing** toolbar to open the **Text tool palette**.

26.　　*Select* the **Label** tool ⌷. The **Label Tool Options** dialog box will appear.

Confirm that the **Placement** option for the labels is set as **Automatically find best placement** for the labels and *change* the **Label Style** to **Choose a style**.

Tool Tidbit

Label ⌷ - This toolbar button allows you to interactively label features in the map display.

Scroll down the **Label Styles** and *select* the Banner style.

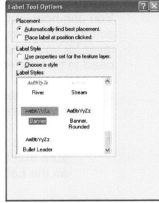

Keep this **Label Tool Options** dialog box open.

27. When you move your mouse over the map display, you should notice that your mouse cursor now looks like a white arrow with the label tool icon attached to it.

Move your **mouse** to one of the selected **school features** and *click* it with your **left mouse button**. A label will be added to the display.

28. Continue to *add* labels to each of the selected schools until all five (5) have labels associated with them. Click Fixed Zoom Out ⤡ until you can view all five labels.

When you have finished, *close* the **Label Tool Options** dialog box.

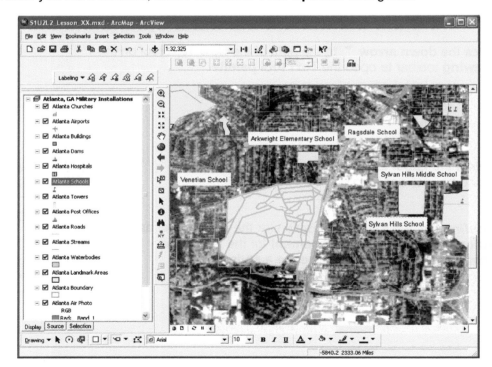

Creating the Map Layout:
In the first lesson of this unit, you created a layout of the map that was used in that project. Because you used that map document as the base for this project, many of the map elements have already been placed on the layout page.

29. **Switch** to **Layout View** by clicking on the **Layout View** button at the bottom of the Data View.

    The page will be set up from the previous lesson.

30. To change the title at the top of the layout page to reflect the content of this map, **double click** the **title** on the layout page. The **Properties** dialog box for the title will open.

    Change the name of the title to **Atlanta, GA Military Installations & Nearby Schools**. **Click** ___OK___ to apply the change and close the dialog box.

    The legend and north arrow located on the layout page will not need to be changed at all. They may need to be resized or moved in order to be visible.

31. The **scale bar** for this map layout needs to be **deleted** and **reinserted**. Although the current scale bar is accurate (ArcMap automatically changed the scale bar when the extent of the data frame changed from Lesson 1 to Lesson 2 when you zoomed in on the selected features), you really should reinsert it so that ArcMap can draw a scale bar that is best for the particular geographic extent being used.

    Click on the **scale bar** on the layout page to select it. **Press** the **Delete** key on your keyboard to delete the scale bar.

    Select Scale Bar... from the **Insert** menu. **Select** the

    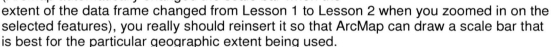

    scale bar style and **click** ___OK___.

    Click and **drag** the new **scale bar** to its appropriate place on the layout page.

32. To edit your **name** and the **date** on the layout page, **double click** the **text box** on the layout page to open **Properties** for the text box. **Change** the date and **click** ___OK___.

Adding the Table to the Layout:
You will now add the list of this list of selected schools to the layout page for this project so that the schools will also be displayed in tabular format.

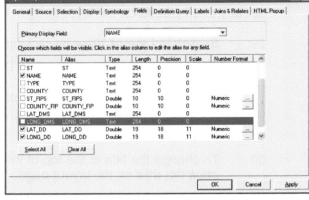

33.   **Open** the **attribute table** for the **Atlanta Schools** data layer. Notice that all of the fields are visible in the attribute table. Only three will be needed in the table that will go on the layout. **Close** the table.

34.   **Double click** the **Atlanta Schools** data layer in the **Table of Contents** to open **Layer Properties**.

        Click the **Fields** tab.

        In the **Visible Fields** box, **deselect** the **FID**, **Shape**, **CODE**, **ST**, **TYPE**, **COUNTY**, **ST_FIPS**, **COUNTY_FIP**, **LAT_DMS** and

        LONG_DMS fields.   **Click** ____OK____ .

35.   **Open** the **attribute table** for the **Atlanta Schools** data layer again. Notice the change to the table fields that are displayed.

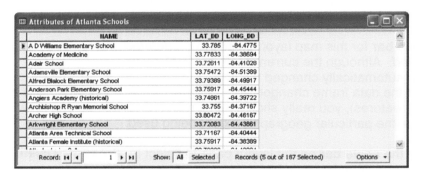

36.   **Click** to display on the Selected features.

37.   **Click** Options ▾ and **select Add Table to Layout**.

        Close the **attribute table**.

38. **Move** the table to a suitable location on the layout page by **clicking** and **dragging** it. You may need to move or resize some of the other map elements to make it all fit nicely on the layout page.

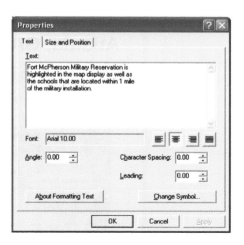

39. To add additional text to the layout page, **select Text** from the **Insert** menu. **Type** the text **Fort McPherson Military Reservation is highlighted in the map display as well as the schools that are located within 1 mile of the military installation**.

    Press Enter on your keyboard. (You may need to type some of the text, press Enter, and then double click the text to open the Properties for the text box and continue to type the text. You will also need to specify **center aligned** text in the Properties dialog box.)

    Click and **drag** the text box above the legend on the layout page.

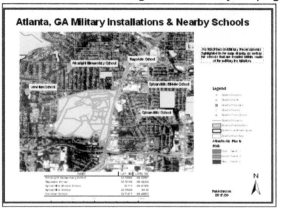

Exporting a Map Layout:

40. To export the map as an image file, **select Export Map...**from the **File** menu.

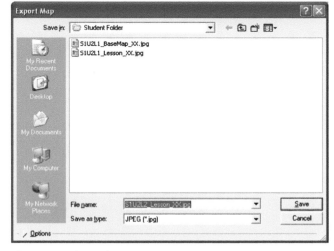

41. **Export** the layout to **your student folder** in **JPEG** format as **S1U2L2_Lesson_XX** (where XX is your initials).

    Click 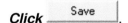 .

Printing a Map Layout:

42. **Select Print...** from the **File** menu to send map to the printer.

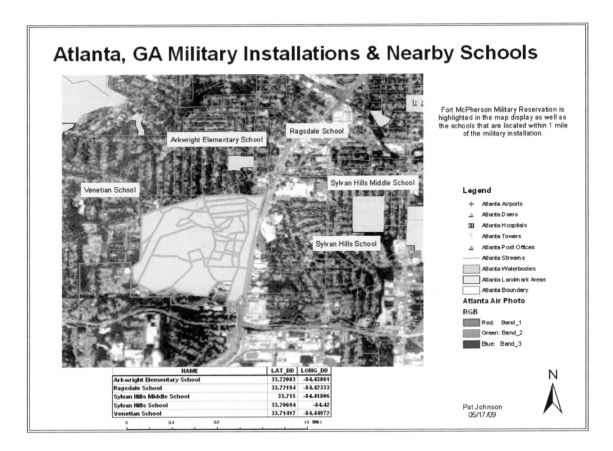

Saving & Closing:

43. **Save** this project.

44. **Select** Exit from the **File** menu or **click** the **Close** ☒ button in the upper right corner of the ArcMap window to exit.

Lesson 2: Managing Geospatial Data Lesson Review

Key Terms
Use the Lesson and the Lesson Activity to define each of the following terms.

1.   **Spatial data**

2.   **Non-spatial data**

3.   **CFCC**

4.   **Source tab**

5.   **Display tab**

6.   **Select by Attributes**

7.   **Select by Location**

8.   **Label tool**

Global Concepts
Use the information from the lesson to answer the following questions. Use complete sentences for your answers.

9.   Why would it be beneficial to join two tables together?

10.   What is the single most important item to consider when joining together two tables?

11.   If you did not want to use Select by Attributes to select a group of features, what other option(s) would you have?

Let's Talk About It...
Answer the following question and share the responses with your instructor and classmates.

12.   Explain how Select by Location is different from Select by Attribute.

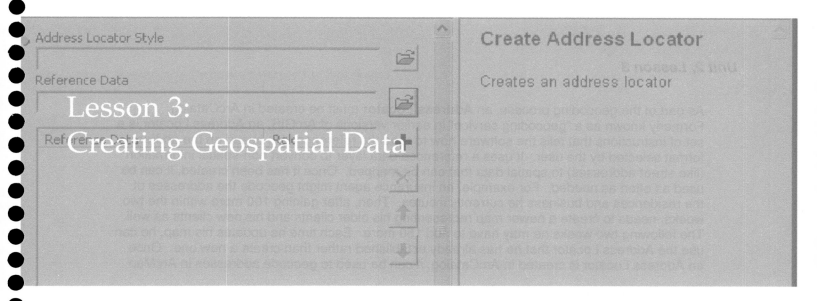

Lesson 3:
Creating Geospatial Data

One of the great benefits of using geospatial technology is that you can cater the information in your study to meet your needs. If you need to study transportation routes, you can create a map showing just that. If you need a map showing small to medium sized cities in a state, you can create it. If you need to create a map of all of the fast food establishments in your city, you can do that. For many maps, the data you need can be downloaded for your immediate use. In some cases, you may have to create your own data or data set to display your findings. This lesson will discuss how to do just that using ArcCatalog and a process called Geocoding.

ArcCatalog

As you studied in earlier lessons, ArcCatalog is the program that is used to preview data and metedata and is used to manage data files. You can easily move files from one location to another, as well as rename files. ArcCatalog is also used for another important function, to create new data layers (or shapefiles). These data layers can be added to ArcMap and can be edited to display the data you create.

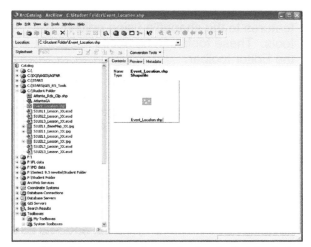

One of the many features of using ArcCatalog is to create new files such as the file shown above.

Geocoding

A common way to find locations is by using street addresses. It is an easier way to describe and report a location relative to a street, neighborhood, or city rather than using an absolute location based x,y coordinates. A GIS however does not display spatial data based on relative location, it requires a more specific, absolute location. **Geocoding** is a process that allows you use a non-spatial data table (table of addresses) and convert that table into a spatial data layer (shapefile) which can be displayed in a GIS by referencing a known spatial data layer (street network shapefile). The nonspatial data table must contain information about the location of the feature to be mapped. For instance, a table of addresses (relative location) can be estimated into absolute location by referencing a shapefile or a city's streets (spatial data layer).

As part of the geocoding process, an **Address Locator** must be created in ArcCatalog. Formerly known as a "geocoding service" in earlier versions of ArcGIS, an Address Locator is a set of instructions that tells the software how to read the referenced file based on the address format selected by the user. It uses a referenced data layer to convert non-spatial information (like street addresses) to spatial data that can be mapped. Once it has been created, it can be used as often as needed. For example, an insurance agent might geocode the addresses of the residences and business he currently insures. Then, after gaining 100 more within the two weeks, needs to create a newer map representing his older clients and his new clients as well. The following two weeks he may have to add 150 more. Each time he updates his map, he can use the Address Locator that he has already established rather than create a new one. Once an Address Locator is created in ArcCatalog, it can be used to geocode addresses in ArcMap.

Creating a New Shapefile
In ArcMap, you can add points, lines, and polygons to a display using the tools on the Draw toolbar. These points, lines, and polygons however are not actual shapefiles and cannot be easily edited. In order to create a new shapefile, you must do so in ArcCatalog. In ArcCatalog, you specify the name and type (point, line, or polygon) for the new shapefile. You can also

specify the coordinate system or allow ArcMap to set one up for you. It is important to note that you only set up the new data file in ArcCatalog. When it is created, the new shapefile is "empty" in terms of features. You must add the new shapefile to ArcMap and then add features to the shapefile using editing tools. These tools can be accessed using the Editor toolbar.

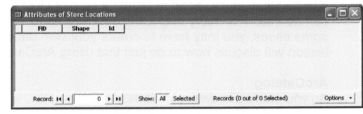

When files are created in ArcCatalog, they do not contain data. Once they are brought into ArcMap, they can be edited and data

The editor toolbar can be utilized for many different applications including creating new features as you will do in the activity for this lesson.

Techniques and Skills
In this particular lesson, you will explore data creation methods using ArcCatalog and ArcMap involving a Homeland Security scenario. Data creation is a valid skill that may be utilized in virtually any scenario such as; police departments keeping track of burglary locations, a city keeping track of its fire hydrants, and real estate agents keeping track of homes for sale. Although in this lesson these tools, techniques and skills are being introduced using a Homeland Security scenario, they can also easily be used for geospatial analysis in other disciplines as well.

Conclusion
ArcCatalog is a program within ArcGIS that helps you manage geospatial data. Among its many uses of ArcCatalog, includes the ability to create data layers (shapefiles). These shapefiles will not include data until you add the data to them in ArcMap using editing tools.

Using an existing reference data set (usually a street network), you can find a single address or multiple addresses and plot the location on the map. You must first create an Address Locator in ArcCatalog. ArcMap will use the Address Locator to find non-spatial locations on the map.

Lesson 3: Creating Geospatial Data

In this lesson, you will continue to explore the various ArcGIS software tools and techniques while performing applications that are relevant to Homeland Security. In this particular lesson, you will find a location in the street network, given only its address. A street dataset is referred to as a street network because it is made up of a system of lines that are interconnected. Attributes contained in the street network dataset are used to match addresses using an address locator that is created in ArcCatalog. This is done by a process called address geocoding. You will gain more experience with geocoding in future lessons. This lesson introduces you to the concept using one address.

Homeland Security Scenario
Officials with a sporting event call police to inform of a mysterious, suspicious package that has been received and they are unsure of its contents. They contacted your office so that appropriate measures could be taken in case the package contained hazardous materials that could endanger the event attendees and the surrounding area.

Finding the Event Location:
You already know that you can use spatial data in a GIS project to create maps. However, what if you want to know where a specific location is on a map? In order to use the street network data layer for a community to locate an address within the network, you must first use ArcCatalog to create an "address locator" or prepare the street network data for use in finding an address.

1.	***Launch* ArcCatalog** by either ***double-clicking*** the shortcut on your computer desktop or by clicking the Windows Start button **start**, ***pointing*** your mouse to **Programs/ArcGIS** and ***selecting* ArcCatalog**. The ArcCatalog window will open.

	Take a minute to refer to the different parts of the ArcCatalog window that were documented in Unit 1 lessons. Make a note of where the toolbars and other components of the ArcCatalog window are located.

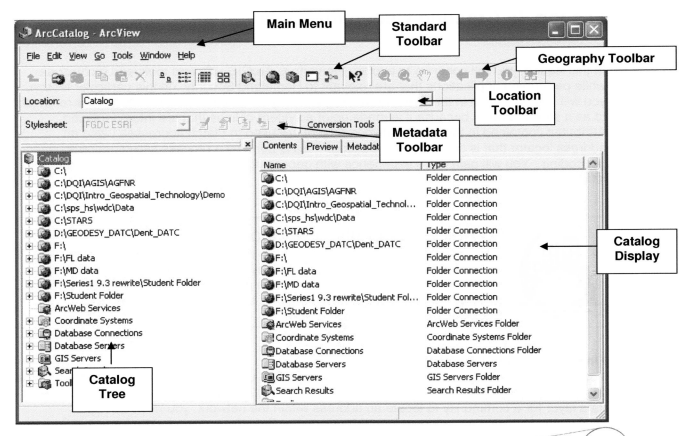

Connecting to a Folder:

2.   In the **Catalog Tree**, *click* the **Connect to Folder** button on the **Standard Toolbar**. **Connect to Folder** allows you to create a "shortcut" to commonly used folders and reducing the amount of time it takes to navigate to find data.

In the **Connect to Folder** dialog box, *navigate* to

C:\STARS\GIS_RS_Tools folder and *click* OK.

The path to the **C:\STARS\ GIS_RS_Tools** folder will appear in the **Table of Contents**.

Tool Tidbit

Connect to Folder - This toolbar button establishes a connection to a commonly used data folder.

3.   Notice that the contents of this folder are listed in the **Catalog Display**. *Double click* the **C:\STARS\GIS_RS_Tools** folder in the **Catalog Tree** so that the individual data files are also listed in the **Catalog Tree**.

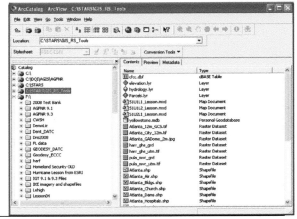

Previewing Data in ArcCatalog:
4. *Click* on the **Atlanta_Rds_Clip.shp** layer in the **Catalog Tree**. In the **Catalog Display**, *click* the **Preview** tab to view the data layer.

   At the bottom of the **Catalog Display**, *change* the **Preview** option from **Geography** to **Table**.

5. *Scroll* to the right in the **Catalog Display** to view the different fields in the table. Notice how the segments of the street network are organized.

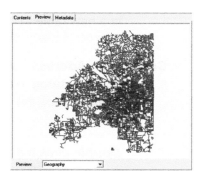

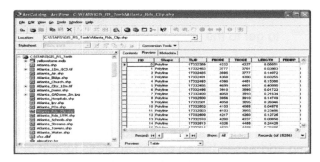

Each street in the network is made up of several separate line segments. The different fields in the attribute table are described as follows:

Field	Description
TLID	TIGER/Line ID
FNODE	"From" node id (*Note: Nodes are the connecting points of the line segments*)
TNODE	"To" node id
LENGTH	Length in miles
FEDIRP	Feature direction prefix (street name prefix) – N, S, E or W
FENAME	Feature name (street name)
FETYPE	Feature type (street type) – Rd., St., Dr., Ave., etc.
FEDIRS	Feature direction suffix (street name suffix) – N, S, E or W
CFCC	Census Feature Classification Code (type of road classification)
FRADDL	"From" address on left side of the street
TOADDL	"To" address on left side of the street
FRADDR	"From" address on right side of the street
TOADDR	"To" address on right side of the street
ZIPL	ZIP code on left side of the street
ZIPR	ZIP code on right side of the street

Whenever ArcMap, or any other application program that gives a location when provided an address with a street network, the program finds the line segment that most closely matches the address provided using the various fields in the street network attribute table. Using the "From" and "To" addresses on the left and right sides of the street segment, the program approximates the location based on the numeric address given.

In order to find the location of a feature using a specific address (or multiple addresses contained in a table), an address locator must be created in ArcCatalog. This process was called creating a geocoding service in earlier versions of ArcGIS and ArcView. An address locator uses specific methods for matching addresses data to a certain reference data layer (in our case, a street network) based on the locator style that is chosen. Once an address locator is created, a "geocoding index" is created and stored in a new table to make it easier to map the addresses without having to search the entire reference layer (street network).

Creating an Address Locator:

6.   In the **Catalog Tree**, *scroll* to the bottom of the tree until you see ⊞ 🗃 Toolboxes . **Double click** the ⊞ 🗃 Toolboxes icon in the **Catalog Tree**.

7.   **Double click** the ⊞ 🗃 System Toolboxes icon and then the ⊞ 🛠 Geocoding Tools .

8.   **Double click** the ⊞ 🛠 Geocoding Tools to get to the Create Address Locator.

9.   **Double click** on 🪓 Create Address Locator to open the **Create Address Locator** dialog box.

There are many different types of address locator styles. An address locator style provides the framework for matching addresses based on the address format of the reference data. Some commonly used address locator styles include:

- **Single Field** – The Single Field address locator style lets you create address locators for address data that contain the location information in a single field.
- **US Streets** – The US Streets address locator style lets you create address locators for common addresses encountered in the United States. One advantage of this address locator style is that it permits you to provide a range of house number values for both sides of a street segment. With this, the address locator cannot only deliver a location along the street segment, but it can also determine the side of the street segment where the address is located.
- **US Alphanumeric Ranges** – The US Alphanumeric Ranges address locator style lets you create address locators for United States addresses that contain alphanumeric house number ranges. Such alphanumeric house ranges are used in some regions of Wisconsin and Illinois. The alphanumeric portion of the address usually represents a grid zone. For example, the address N84W 16301 W Donald Ave suggests that the address is not only at 16301 W Donald Avenue, but that it is also in grid zone N84W.
- **US Hyphenated Ranges** – The US Hyphenated Ranges address locator style lets you create address locators for United States addresses that contain hyphenated house number ranges. The hyphenated ranges depict a number that is usually the number of the cross street, followed by a hyphen, then the actual number of the house along the street (for example, 76-20 34th Ave). One location that uses this type of address style is Queens, New York. The first number indicates either the north or west cross street. The second number indicates where on the block the building is located.
- **US One Range** – The US One Range address locator style lets you create an address locator for street segments with address ranges. This address locator style is similar to the

US Addresses address locator; however, this style requires only one range for each road segment.

- **US One Address** – The US One Address locator style lets you create address locators for United States addresses. US One Address locators use feature classes with polygon or point geometry as reference data. Each feature in the reference data corresponds to a single address.
- **US Cities with State** – The US Cities with State address locator style lets you create address locators for city names that contain fields that have city and state name information.
- **ZIP 5 Digit** – The ZIP 5 Digit address locator style lets you create address locators for postal codes. (*Source: Environmental Systems Research Institute (ESRI), http://webhelp.esri.com/ arcgisdesktop/9.1/body.cfm?tocVisable=1&ID=1643&TopicName=Commonly%20used%20ad dress%20locator%20styles*)

10. **Select** the **US Streets** address locator style in the **Create New Address Locator** dialog box.

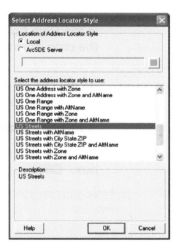

Click ___ OK ___ . The **New US Streets Address Locator** dialog box will appear.

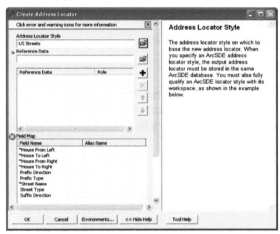

11. In the **Reference Data** box, *click* the **Browse** button

to select the reference data layer. *Navigate* to the **C:\STARS\GIS_RS_Tools** folder and *select* the **Atlanta_Rds_Clip.shp** data layer.

Click **Add** . The fields that contain the necessary data to locate addresses will automatically be filled in to the appropriate boxes associated with the primary reference table.

When you first add this layer, you will get two ⊗ indicating the data set is not ready for use. To clarify the use of the data set, you will need to assign a role to it. *Single click* in the Role box to the right of the data set you just added. *Select* **Primary table** from the drop down list.

In the Output Address Locator box, select the and browse to your student folder. *Name* the **Output Address Locator** as **AtlantaGA**. *Click* **Save** return back to the **Create Address Locator** dialog box. *Click* **OK** to run the locator.

Once the index is created for this new address locator, the address locator appears in the **Catalog Tree**.

Click **Close** to close the box.

You will now use the map document you saved from Lesson 2 for the activities contained in this lesson.

12. *Start ArcMap* by *clicking* the **Launch ArcMap** button 🌐 in **ArcCatalog**. When **ArcMap** opens, *select* to open ⦿ An existing map: . *Open* the **S1U2L2_Lesson_XX.mxd** file in **your student folder**.

The map document will open and will still be in **Layout View** from the last lesson.

Tool Tidbit

Launch ArcMap 🌐 - This toolbar button launches **ArcMap** from **ArcCatalog**.

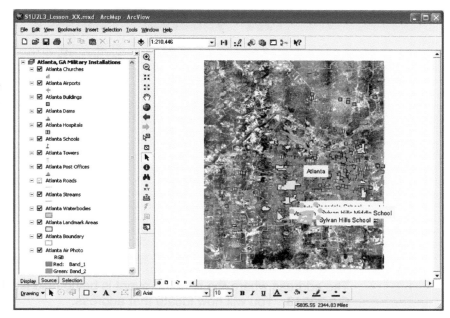

13.   **Click** 🌐 to switch back to **Data View**. **Click** the **Full Extent** button 🌐 to zoom out to the entire geographic extent of the project.

14.   **Select** **Save As...** from the **File** menu and **save** this map document as

S1U2L3_Lesson_XX.mxd (where **XX** is your initials) in your student folder.

15.   **Open** **Layer Properties** for the **Atlanta Roads** data layer. In the **General** tab, **remove** the **scale threshold** from the layer by selecting to ⦿ Show layer at all scales . **Click** OK . The roads will now display on the map.

16.   **Single click** on the names of the **schools** listed in the map display and **delete** the name boxes. To remove the highlighted features, **single click** on the **deselect** button 🔲 on the tools toolbar.

17.   **Double click** the **Atlanta, GA Military Installations** data frame and **change** the **name** of the data frame to **Atlanta, GA Haz-Mat Threat**.

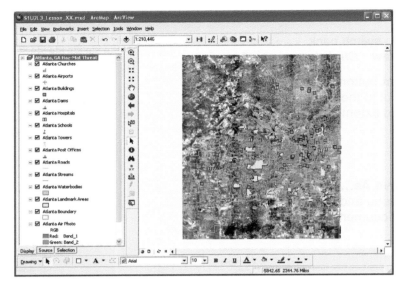

The address where the sporting event is being held is: **1 Georgia Dome Drive, Atlanta, GA 30313**. You will now locate that address.

Finding an Address:

18. To find this location, **click** the **Find** button. To find an address, **click** the **Addresses** tab.

   You will now specify the **address locator** that you just created to be used to find the address.

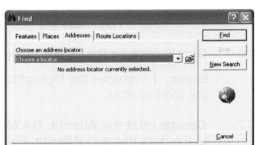

19. **Click** the **Browse** button. The **Add Address Locator** dialog box will appear.

   Navigate to your **student folder** and **select** the **AtlantaGA** address locator that you created.

Tool Tidbit

Find - This toolbar button allows you to find features on a map or find addresses in a street network.

Introduction to GIS Tools and Processes, version 7

The address locator will now be displayed in the **Find** dialog box.

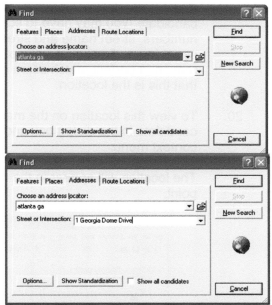

Click in the **Street or Intersection** box and *type* the address **1 Georgia Dome Drive**.

Click [Find]. Any results of the search for **1 Georgia Dome Drive** will appear at the bottom of the **Find** dialog box. This search resulted in one (1) candidate.

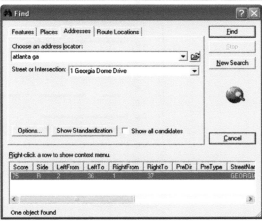

Results are scored based on how well the address matched the reference table. Notice that this candidate scored 75 of a possible 100 points. If you look at the fields for this candidate (you may have to scroll to the right to see them all), notice that the address numbers, street name and street type match. However, in the roads reference table, the street has a directional suffix of NW that was not included in our address. For this reason, a perfect match score of 100 cannot be made. However, you can feel confident that this is the location.

20.   To view this location on the map display, *right click* the candidate in the dialog box and *select* **Add Point** from the context menu.

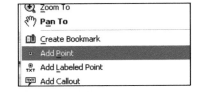

The location will appear in the map display as a large black point.

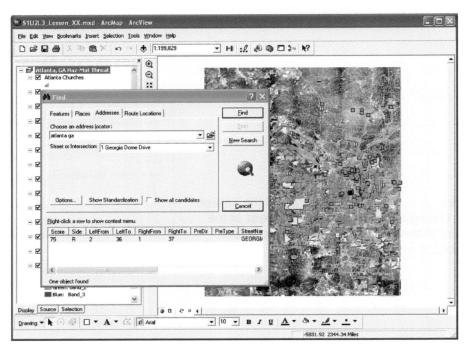

Close the **Find** dialog box.

Even though this graphic has been added to the map, it is not included as a data layer for the project. To make a data layer containing this location so that it can be used for analysis purposes, you must create a new data layer in ArcCatalog and then use ArcMap editing tools to add the feature to the new data layer.

Creating a New Shapefile Data Layer:
In order to perform any future spatial analysis pertaining to this location, this point must be stored as a data layer. Even though a graphic symbol is added to the map display, you cannot use this point to perform any analysis. In order to do that, you must create a new data layer that contains a feature (a point) that represents this location. To create a new data layer (in this case, a shapefile), you must use ArcCatalog.

21. Make **ArcCatalog** active again. *Click* the **Connect to Folder** button  and *navigate* to your **student folder** in the **Connect to Folder** dialog box in order to have the connection to your student folder listed in the Catalog Tree so you will not have to navigate to it each time you use it in ArcCatalog.

22. In the **Catalog Tree**, *right click* your **student folder** and *select* **New ▶ Shapefile...** The **Create New Shapefile** dialog box will appear.

23. *Name* the new shapefile **Event_Location** with a **Point** feature type.

    This new shapefile does not have a spatial reference yet. Spatial reference refers to the coordinate system and any projection that is assigned to geographic data. It is not required to specify a spatial reference when a new data layer is created. However, you will specify the spatial reference.

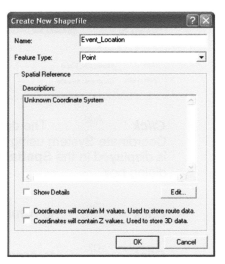

Click Edit... to open the **Spatial Reference Properties** dialog box.

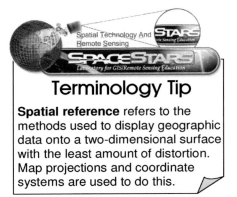

Terminology Tip

Spatial reference refers to the methods used to display geographic data onto a two-dimensional surface with the least amount of distortion. Map projections and coordinate systems are used to do this.

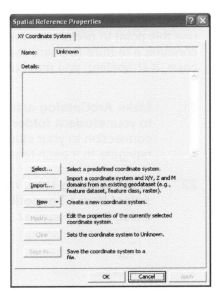

At this point, you can either select the coordinate system yourself or use the coordinate system of another data layer by importing that data layer's spatial reference information.

Click Import... to specify the spatial reference of another data layer.

Navigate to the **C:\STARS\GIS_RS_Tools** folder and ***select*** the **Atlanta_City_12m.tif** image layer

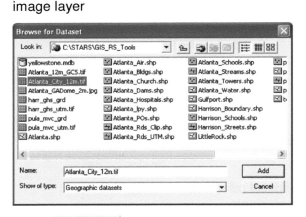

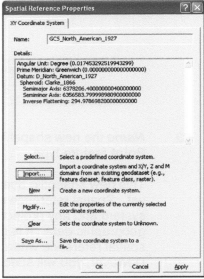

Click Add . The details for the Geographic Coordinate System using North American Datum 1927 is displayed in the **Spatial Reference Properties** dialog box.

Click OK to ***return*** to the **Create New Shapefile** dialog box.

Click OK to ***create*** the **new shapefile**. The new shapefile will be added to your **student folder**.

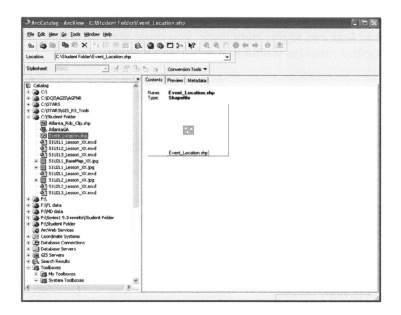

Adding Features to a New Shapefile Data Layer (Heads-Up Digitizing):

Keep in mind that the shapefile that you just created does not yet contain any data. You will now add data to this new shapefile through a process called **heads-up digitizing**. **Heads up digitizing** involves using ArcMap editor drawing tools with your mouse to add features to a data layer. In this activity, you will heads-up digitize a point feature but, you can also perform heads-up digitizing for line and polygon data layers.

24. **Click** the **ArcMap** S1U2L3_XX.mxd - ... icon in the taskbar to make ArcMap active again.

> ## Terminology Tip
> **Heads-up digitizing** is the process of creating geographic data features using computer drawing tools and a mouse. Special digitizing tablets can also be used to create geographic feature data.

25. **Add** the **Event_Location.shp** data layer that you just created from your **student folder**.

When the layer is added to ArcMap, no features will display in the map for this layer because no features have been added to this layer yet.

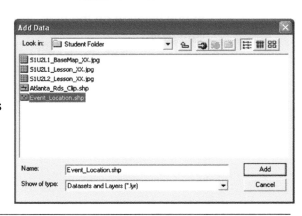

In order to add features to this data layer, you must start an editing session. To do this, the **Editor** toolbar must be visible in the ArcMap window.

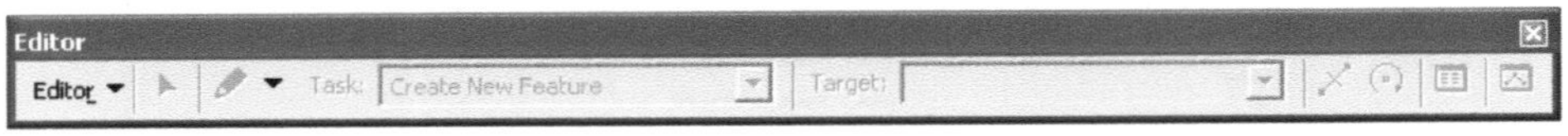

26. If the Editor toolbar is not currently visible in the window, ***select*** **Toolbars ▸ Editor** from the **View** menu.

    When the toolbar is added, ***select*** **Start Editing** from the Editor ▾ drop-down menu. The **Start Editing** dialog box will appear to prompt you for the folder location of the data layer that you want to edit.

    Select your **student folder** and ***click*** OK.

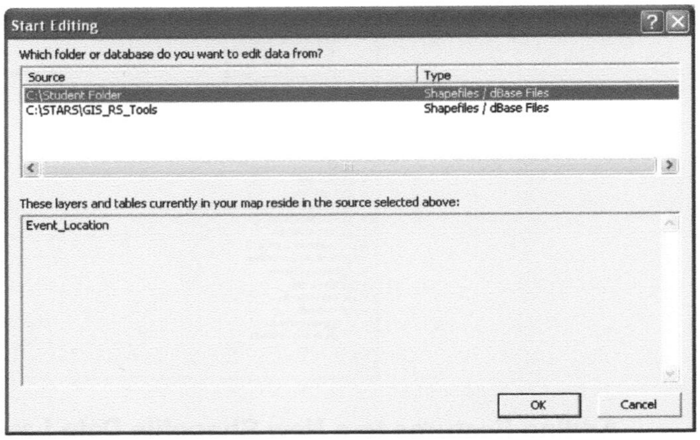

27. In the **Editor** toolbar, make sure that **Create New Feature** is set as the **Task** (meaning what you want to do in this editing session) and **Event_Location** is set as the **Target** (meaning that this is the data layer that you want to edit).

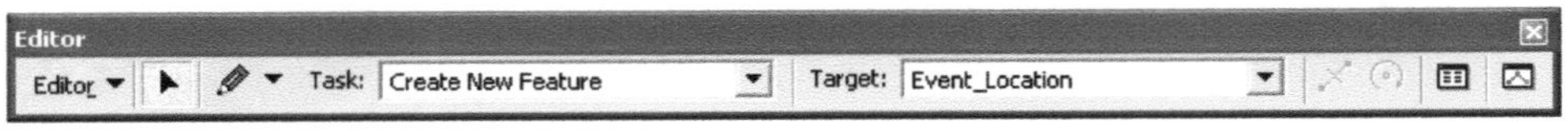

Click the **Sketch** tool  to draw features in this data layer.

28. With your **left** mouse button, ***click*** on the **black point graphic** that was added to the map to mark the location of the event.

    When you do this, a point will be added to the data layer at that location on the map and it will be highlighted in light blue.

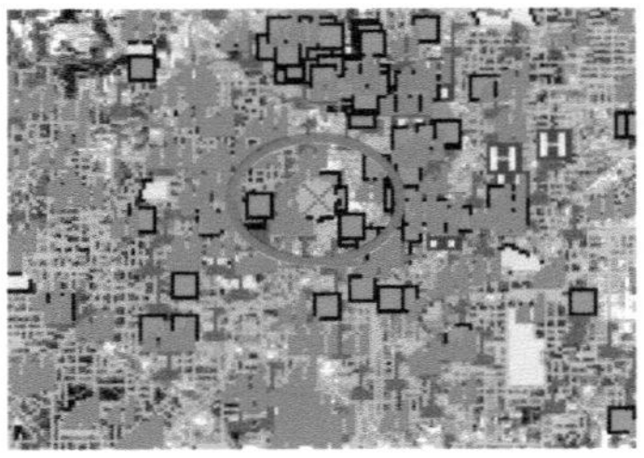

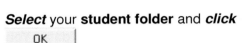

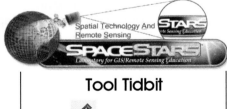

Tool Tidbit

Sketch 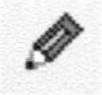 - This Editor toolbar button allows you to create new geographic features in a data layer.

29.   To end this editing session, **select Stop Editing** from the Editor ▼ toolbar drop-down menu.

30.   When prompted to save the edits, **click** Yes .

31.   To remove the **black point graphic** that was placed on the map originally, **click** on the **point** with your mouse and **press** the **Delete** key on your keyboard. The point will disappear and you will be left with the symbol for the geographic feature that you just added.

Editing Feature Symbology from the Table of Contents:
You have already changed the symbols of data layers using the Layer Properties dialog box under the Symbology tab. You can also edit symbology from the Table of Contents. You will gain experience doing that now.

32.   **Click** on the **symbol** for the **Event_Location** data layer in the **Table of Contents**. The **Symbol Selector** dialog box will appear.

Click More Symbols ▼ and **select** **Crime Analysis** to view additional symbols.

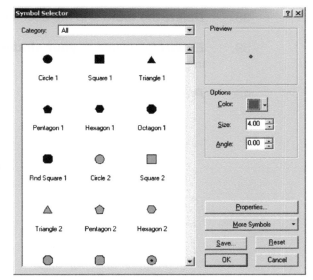

Select the Hazardous Materials symbol from the list and **click** OK . The symbol will be changed.

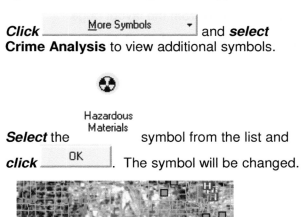

33.   **Save** this project.

Creating the Map Layout:

34. *Switch* to **Layout View** by clicking on the **Layout View** button at the bottom of the Data View.

     The page will be set up from the previous lesson.

35. To change the title at the top of the layout page to reflect the content of this map, *double click* the **title** on the layout page. The **Properties** dialog box for the title will open.

     Change the name of the title to **Atlanta, GA Haz-Mat Threat**. *Click* OK to apply the change and close the dialog box.

     The legend and north arrow located on the layout page will not need to be changed at all. (If you accidentally delete either of these items, you can always reinsert it by using the options in the **Insert** menu.)

36. *Delete* and *reinsert* the **scale bar** for this map layout.

     Click and *drag* the new **scale bar** to its appropriate place on the layout page.

37. *Edit* the date in the text box containing your name and date.

38. *Delete* the **table** and the accompanying **text box** from the layout page.

39. *Export* the layout to **your student folder** in **JPEG** format as **S1U2L3_XX** (where XX is your initials).

40. *Print* the map.

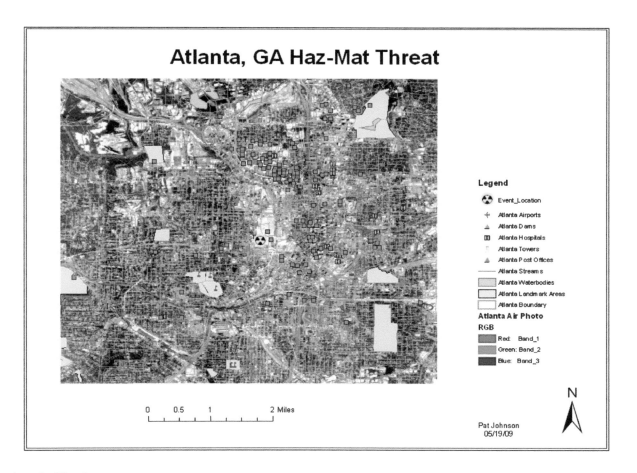

Saving & Closing:

41.   **Save** this project.

42. **Select** Exit from the **File** menu or **click** the **Close** button in the upper right corner of the ArcMap window to exit.

Lesson 3: Creating Geospatial Data Lesson Review

<u>**Key Terms**</u>
Use the Lesson and the Lesson Activity to define each of the following terms.

1.   **Geocoding -**

2.   **Address Locator** -

3.   **Connect to Folder** –

4.   **Find button** –

5.   **Spatial Reference** -

6.   **Heads-up Digitizing** -

7.   **Sketch** –

<u>**Global Concepts**</u>
Use the information from the lesson to answer the following questions. Use complete sentences for your answers.

8.   After creating a file in ArcCatalog, how can you add data to it?

9.   Why is it beneficial to use Connect to Folder?

10.   Is using the Find Tool the same as Geocoding in finding an address? Explain your answer.

<u>**Let's Talk About It...**</u>
Answer the following question and share the responses with your instructor and classmates.

11.   In this Lesson and in the Lesson Activity, you will learn how to create a data layer in ArcCatalog and ArcMap using a Homeland Security scenario. Name at least three other scenarios for data might have to be created.

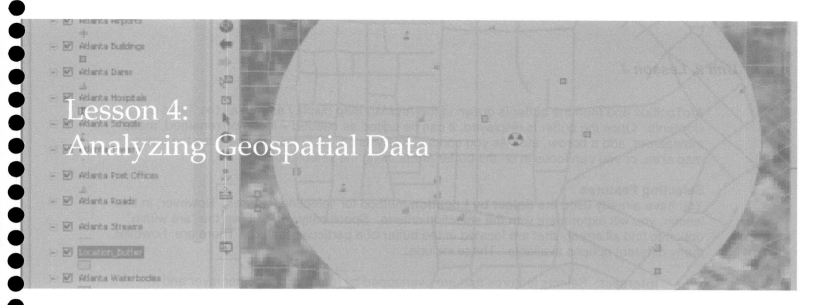

At this point you have had experience with two ArcGIS programs where data can be created, displayed, edited, and much more. There is another ArcGIS program that you have been introduced to, but not worked with much yet, ArcToolbox. ArcToolbox must be operated within one of the other programs. To activate

ArcToolbox, select the  to show the ArcToolbox window. The window will open in the ArcGIS application you are running and contains many sets of tools which can be used to perform various functions such as establishing buffers, clipping layers, joining layers, conversions, geocoding, image analysis, and statistical analysis. Additional toolboxes can be created to perform specific analyses created by the user, however that skill will be taught in a later lesson.

ArcToolbox manages these many processing and analysis functions and provides an organized listing of the more common analysis and processing tasks. All of these tools are organized into toolboxes and toolsets based on the type of analysis that is performed.

Buffers

In this particular lesson, you will use ArcToolbox to perform buffer analysis for a data layer. **Buffers** are polygon features that are drawn at a specified distance around features in a data layer. You can specify to draw the buffers around all features or only selected features. A large scale example would be drawing a buffer around the sprinkler heads in a yard to determine which parts of the lawn will receive water and where they could be placed to cover the entire yard. A smaller scale example would be to place a 700 yard buffer around an area in which a building was going to be demolished. This would establish a safe vantage point without risk of injury to those who wish to watch.

ArcToolbox provides a streamlined one-step buffer process. You can perform the buffer processing using

This map shows a 700 yard buffer around a building that will soon be demolished.

ArcToolbox and then the buffer is drawn in the ArcMap map display as well as the Table of Contents. Once the buffer is displayed, it can be edited as needed – it can be renamed, made transparent, add a border, etc. As you conduct analysis on your map, you can use the entire map area, or you can focus in on the buffer areas using the select tools.

Selecting Features

You have already used the **Select by Location** method for selecting features. However, in this lesson, you will experiment with the selection criteria. Specifically you will use the "are within" option to find all streets that are located in the buffer of a particular feature. There are, however, many different options available. These include:

- **Intersect:** Selects features that are overlapped by features of another layer and features that border the features of the other layer
- **Are within a distance of:** Selects features that are located within the distance from the features that are selected in a data layer
- **Contain:** Selects features in one layer that contain the features of another (and boundaries of the features ARE allowed to touch)
- **Completely Contain:** Selects <u>polygons</u> in a layer that completely contain features from another layer
- **Are Within:** Selects features in a layer that are inside the <u>polygons</u> of another layer
- **Are Completely Within:** Selects features in a layer that are completely inside the <u>polygons</u> of another layer
- **Are Identical to:** Selects any feature that has the same geometry as a feature of another layer (feature types must be the same – point, line or polygon)
- **Touch the Boundary of:** Selects lines and polygons that share line segments, vertices, or end-points with the lines in the layer (but features are not selected if they cross the lines in the layer)
- **Share a Line Segment with:** Selects <u>line</u> and <u>polygon</u> features that share line segments with other features
- **Are Crossed by the Outline of:** Selects features that are overlapped by features of another layer
- **Have Their Centroid in:** Selects <u>polygon</u> features in one layer that have their center in <u>polygon</u> features of another layer

As you continue to work with ArcGIS tools and techniques, you will become familiar with many of these listed.

Up to this point in the various analyses that you have conducted, all of your selections that you have performed have created a newly selected group. This lesson takes your selection skills to the next level in that you will now perform analyses on the group that you initially selected. In effect, you will perform a selection on the group that you have already selected, thus narrowing down the number of entries you have. For example, if you had selected a group of houses within a buffer zone, you can run a more specific selection process to specify to select only the two story

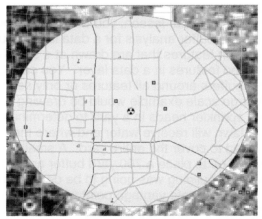

Now that only the streets within the buffer are selected, you can specify to remove streets within this selection that have four lanes.

houses within this zone. After the second selection has been run, only the two story homes would be highlighted because they meet the new criterion.

In addition to selecting from a current selection, you can also specify to remove features that meet a certain criterion or set of criteria from the currently selection. If the new criterion required that you remove any houses from the selection that have swimming pools, a remove features selection could be run.

Adding Fields to Attribute Tables

When you download data from the Internet or when you create your own data sets, there will be times when you need to add new fields to attribute tables to customize them for your study. There are two ways that you can do this – one using ArcCatalog and one using ArcMap. In this particular lesson, you will use the method that utilizes ArcCatalog. To do this, you right click a data layer in the Catalog Tree and open Properties. In the Fields tab of the Properties dialog box, you can enter the fieldname and data type for the new field using up to ten characters. The new field will immediately be added to the attribute table for that layer. You can then begin an editing session in ArcMap and add values in the new field. Once the new data is added, you may have the need to display these new features using unique values.

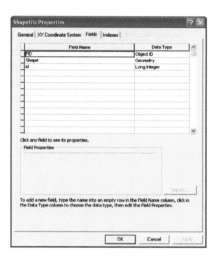

Symbolizing by Unique Value

Up until this point, you have primarily used the Single Symbol method of showing data in a map display. In this lesson, you will use the Unique Values method. In order to use the Unique Value method, you must specify a value field from the data layer. This method allows you to show a unique symbol for each unique value in the value field. This is typically not the best method to use for quantitative (numeric) data but works well for textual data that has only a specified number of possible values. There are other ways of displaying numerical data that will be explored in later lessons.

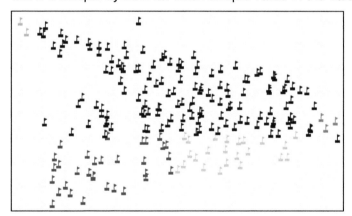

These schools are categorized by unique values based on their location within the county. Can you think of other ways that schools could be categorized?

Data Inventory

In this particular lesson, you will explore advanced selection methods using Select by Attribute and Select by Location. In addition, you will perform a buffer analysis and edit data by adding a new field to a data layer's attribute table. You will explore all of these tools involving a Homeland Security scenario. Keep in mind that, although these tools, techniques and skills are being introduced using Homeland Security scenarios, they can also easily be used for geospatial analysis in other disciplines as well.

Conclusion

The ArcToolbox window contains commonly-used geoprocessing tools and additional toolboxes can be added by the user as needed. Buffers, which are helpful in determining a specified distance from a area or group of areas, can be created using the ArcToolbox Proximity tool. Once these area(s) are determined, it is sometimes useful to perform analysis within the selected areas. Different selection methods can be used with the Select by Location and Select by Attribute techniques to control how and what features are selected.

One of the many editing techniques that can be performed on layer attribute tables is to add new fields. This can allow a user to customize the data set for their specific needs. New fields can be added to layer attribute tables by editing Shapefile Properties in ArcCatalog or they can be added in ArcMap. Once data layer attribute tables have been established, layer features can be symbolized by unique value to differentiate between different types of features within the same data layer. There are other ways to symbolize data in layers, these other methods will be covered in future lessons.

Lesson 4: Analyzing Geospatial Data Lesson Activity

In the last lesson, you mapped the location of an address in Atlanta. Because you created a new shapefile data layer to contain a feature to mark that location, you can now perform various types of analysis with that location.

In Unit 1 you were introduced to **ArcToolbox** as a source for common geoprocessing tools. These ArcToolbox tools perform many processing functions that can also be performed in other parts of ArcGIS; however, ArcToolbox conveniently organizes them by processing category into toolboxes.

In this lesson, you will continue to gain experience with ArcGIS tools and techniques while addressing scenarios pertinent to the field of Homeland Security. This particular lesson introduces you to the concept of buffering features.

Homeland Security Scenario

Upon further review of the package found at the sporting event, hazardous materials investigators have been unable to determine the substance contained in the box. To err on the side of caution, first responders have decided to evacuate the stadium and cordon off all incoming traffic up to one-half (½) mile from the stadium.

Establishing a Half-Mile Buffer:
In order to quickly evacuate the people at the event site, it is important to establish a buffer zone around the location (that has already been determined to be a ½ -mile buffer) and then control traffic so that people located inside the buffer area can evacuate and people outside the buffer area will be prevented from entering the buffer zone.

To create a buffer, you will use **ArcToolbox**. ArcToolbox, as you may remember from Unit 1, is the ArcGIS component that provides geoprocessing tools that can be used in ArcMap or performed in ArcCatalog. You can launch ArcToolbox directly from ArcMap or ArcCatalog.

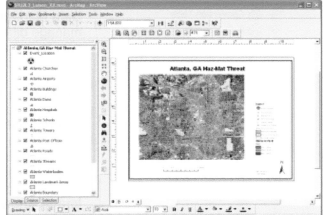

You will use the map document you saved from Lesson 3 for the activities contained in this lesson.

1. *Launch* **ArcMap** and *open* the **S1U2L3_Lesson_XX.mxd** map document you created in the last lesson.

2. *Switch* to **Data View**.

3. *Select* **Save As...** from the **File** menu and *save* this map document as **S1U2L4_Lesson_XX.mxd** (where **XX** is your initials) in your student folder.

---

4. *Click* the **Show/Hide ArcToolbox Window** button

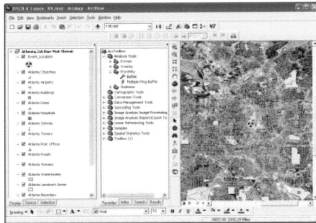 to display the **ArcToolbox** geoprocessing tools.

Tool Tidbit

Show/Hide ArcToolbox

Window  - This toolbar button allows you to access the ArcToolbox geoprocessing tools

Take a minute to notice the different geoprocessing tools that are included in the ArcToolbox window. Geoprocessing generally means performing analyses on geographic data. These tools are organized in ArcToolbox. Many of these will be covered in a later course.

Analysis Tools – contains toolsets with tools to perform geoprocessing techniques on vector data; analyses include clipping (Extract tool), intersections (Overlay tool), buffers (Proximity tool), summary statistics (Statistics tool) and others

Cartography Tools – contains toolsets to allow you to convert a map to cartographic standards

Conversion Tools – contains toolsets with tools to allow you to convert from one data type to another; conversion options include Raster to Vector, table to database, Feature class to geodatabase, Feature to Raster, Feature class to Shapefile and others

Data Management Tools – contains toolsets with tools to manage different types of datasets including feature classes, data layers and raster data; tools include Add XY Coordinates (Features tool), Add Field (Fields tool), Add Join (Joins tool), and others

Geocoding Tools – contains tools to allow geocoding of addresses using a reference data layer

Image Analysis Image Processing Tools – contains toolsets with tools to perform various processes to images; processes such as Classification (Supervised and Unsupervised), Spatial Enhancement (Resolution Merge), Spectral Enhancement (Vegetative), and more.

Image Analysis Import-Export Tools – contains tools that allow for images to be imported and exported using specified formats

Linear Referencing Tools – contains tools that more accurately model the dynamic nature of linear feature datasets

Samples Tools – contains tools for common analysis, conversion and data management techniques

Spatial Statistics Tools – contains tools that perform various statistical analyses in using different means; processes such as Analyzing Patterns (Average Nearest Neighbor),

Mapping Clusters (Hot Spot Analysis), Measuring Geographic Distributions (Central Feature), and more.

Keep in mind that some of these techniques can be performed without use of ArcToolbox – such as geocoding, joining, adding fields to tables and others. However, ArcToolbox conveniently organizes all of these tools and techniques by the type of process that it is.

To create a buffer around the event site that you located in the last lesson, you will use the **Buffer tool** located in the **Proximity toolset** in the **Analysis Tools toolbox**.

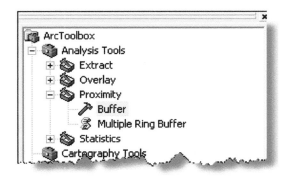

Terminology Tip

Buffers are created to identify a certain specified distance around a feature or features. Example: You can identify the area around a wetland that cannot be commercially developed.

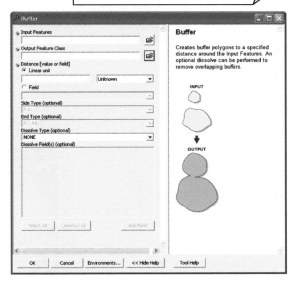

5. In the **ArcToolbox** window, ***double click* Analysis Tools** to open that toolbox. ***Double click*** the **Proximity** toolset to open it. You will see the **Buffer** tool.

6. ***Double click*** the **Buffer** tool to open the **Buffer** dialog box.

7. ***Click*** the **Browse** button 📂 to the right of the **Input Features** box. **Input Features** is the data layer that you want to place a buffer around.

   Navigate to your **student folder** and ***select*** the **Event_Location.shp** layer that you created in the last lesson.

In the **Output Feature Class** box, *name* the output layer **Location_Buffer.shp** in your **student folder**. The **Output Feature Class** is the data layer that will be created containing the buffer.

In the **Distance** section, *type* .5 in the **Linear unit** box and *select* **Miles** from the drop-down list of units of measure. This will set up the half mile buffer. The other options do not need to be changed.

When you have finished, *click*

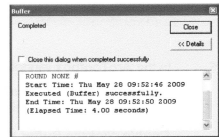

to create the buffer.

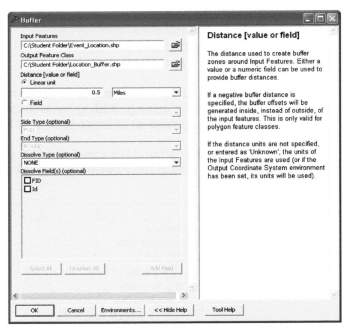

A dialog box showing the progress of creating the buffer will appear.

Click **Close** when the process is complete and the buffer appears in the map display.

Right click on the **Location_Buffer** layer in the Table of Contents and *select* **Zoom to Layer**.

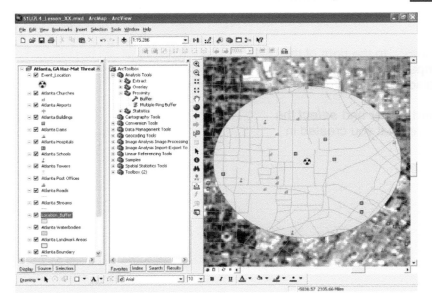

Knowledge Knugget

Why isn't the buffer a circle? Although the distance is a true half mile from the center of the buffer to its edges, the buffer appears to have an ellipsoid shape. This is due to the distortion that occurs when displaying our 3D earth in a 2D display.

8.   *Click* ⊠ to **close** the **ArcToolbox** window.

Selecting Features within the Buffer:
In order to identify the streets located within the buffer zone that will require supervision during the evacuation and until the problem has been finalized should be identified in terms of what type of supervision will be necessary. You will use the **Select by Location** technique to identify the streets.

9.   ***Choose*** Select by Location from the **Selection** menu.

Specify to **select features from** the **Atlanta Roads** data layer that **are within** the **Location Buffer** that you just created.

Click Apply and those streets located within the half mile buffer zone will be selected.

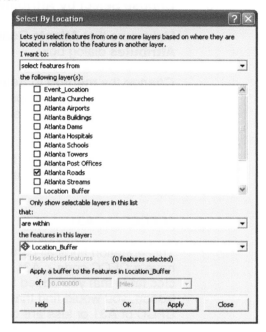

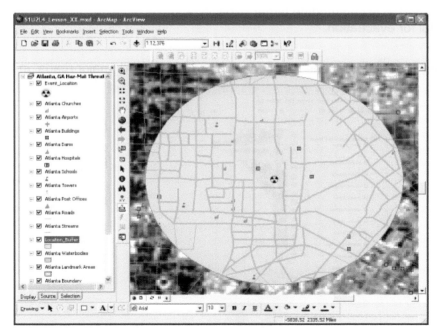

Selecting from a Current Selection:
During an evacuation, major roads are best to use to make traffic flow faster. In this case, larger roads around the event location will be identified and changed to one-way flowing out of the buffer zone to get event attendees and others in the buffer zone away from the site. In order to determine the major roads located in the buffer zone, you will use the **Select by Attribute** technique to select features from the current selection.

10.   *Choose* **Select by Attribute** from the **Selection** menu.

Make sure that **Atlanta Roads** is selected as the **layer** at the top of the **Select by Attributes** dialog box.

Change the **method** to **Select from current selection**. This option specifies to only select features that are already selected if they meet the given criterion.

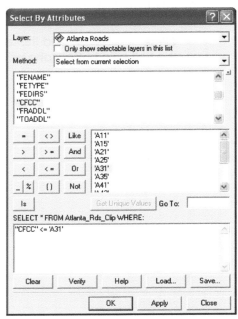

In the query box, *use* the fieldnames and mathematical operator buttons to build the equation:

"CFCC" <= 'A31'

This equation finds all currently selected roads that have a CFCC code ≤ A31. If you were to open the cfcc table that you added to the project in an earlier lesson, you would find that roads that have a CFCC of A31 or less are considered major roads. The roads less than A31 will be roads that are designed to accommodate larger capacity and useful for evacuation.

Click OK to see the roads that are considered "major roads" in the buffer zone.

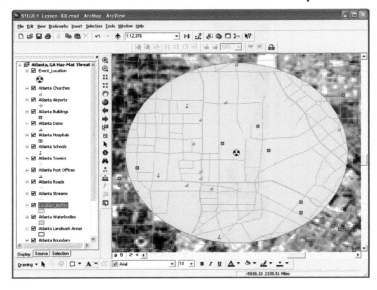

Two roads that are selected from this query are not located in a place that will be beneficial to the evacuation effort. These roads are **Marietta** Street (located in the far eastern portion of the buffer zone) and **Peters** Street (southernmost street segment). You can use the **Select by Attributes** technique to remove these features from the selection.

11. ***Choose*** **Select by Attribute** from the **Selection** menu.

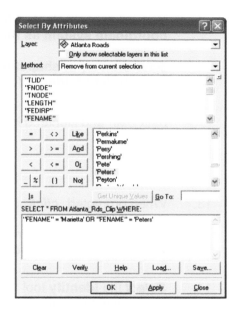

Click **Clear** to clear the previous select query.

This time, ***select*** **Remove from current selection** as the method of selection.

Build an equation using the fieldnames and the mathematical operator buttons to find these two streets.
Click the **Get Unique Values** button to list all of the street names to make building the equation easier.

The equation should state:

"FENAME" = 'Marietta' OR "FENAME" = 'Peters'

Click **OK** to remove these two streets from the selection.

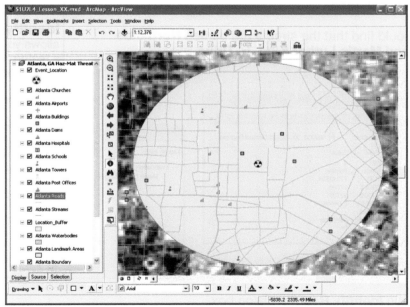

Now that these major roads have been identified, you can export the selected roads as a separate data layer.

Exporting Selected Features as New Data Layer:
Sometimes it is helpful to save specific features from an existing data layer as a separate data layer. In this case, the evacuation routes that have been identified can be exported as a new shapefile.

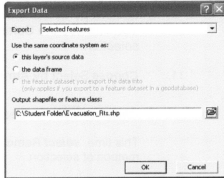

12.   ***Right click*** the **Atlanta Roads** layer in the **Table of Contents** and *select* **Data ▸ Export Data...** from the context menu. The **Export Data** dialog box will appear. You will export **selected features** using the **same coordinate system as the layer's source data**.

        Save the **output shapefile** as **Evacuation_Rts.shp** in your **student folder**. (You can click the **Browse** button

        📂 to navigate to your student folder or simply type the data path and filename into the output shapefile box.)

        Click ▢ OK .

13.   When prompted, ***click*** ▢ Yes to add the exported data as a map layer.

        You will now select the remaining roads within the buffer zone and export those roads to a new shapefile.

14.   ***Select*** the **Identify** tool 🛈 and ***click*** a segment of each of the selected roads in the map display to determine the name of each road.

        You should find that the streets are named **Northside Drive** and **Martin Luther King Jr. Drive**.

> **Tool Tidbit**
>
> **Identify** 🛈 - This toolbar button allows you to view the attributes of a feature by clicking on the feature in the map display.

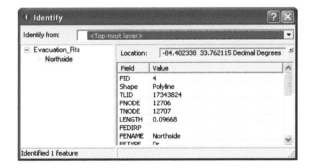

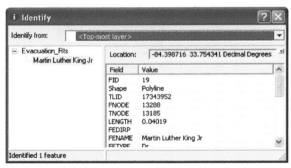

        Close the **Identify Results** dialog box.

---

15. Use the **Select by Location** method to again select those roads that are located in the buffer zone.

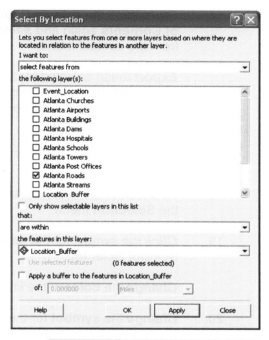

16. Use the **Select by Attributes** method to **remove Martin Luther King Jr.** and **Northside** Drives **from the current selection**. (Be careful because in the list of unique values, there is also a Martin L King Jr. Drive.)

The remainder of the roads inside the buffer zone will be selected.

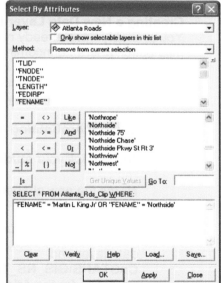

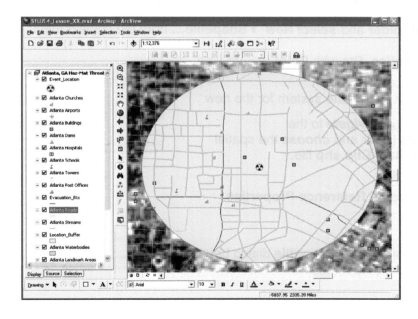

17. ***Right click*** the **Atlanta Roads** data layer in the **Table of Contents** again and *select* **Data ▸ Export Data...** from the context menu.

    Export these selected features to your **student folder** as **Closed_Rds.shp**. These roads will be closed to incoming traffic while there is a crisis situation.

    When prompted, ***click*** Yes to add the new data layer to the map display.

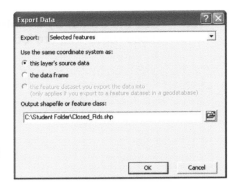

18. To deselect the selected **Atlanta Roads** features, *select* **Clear Selected Features** from the **Selection** menu.

19. ***Click*** the symbol for the new **Evacuation_Rts** layer in the Table of Contents to open the **Symbol Selector**.

    Change the **color** of the line to Red with a **line width** of **2**.

20. ***Change*** the **symbol** for the **Closed_Rds** layer to Orange with a **line width** of **1.25**.

Creating a New Data Layer to Show Police Access Points:
You probably have already noticed that there are several streets that will be affected by a half mile buffer where no incoming traffic will be allowed. Police officers will need to be stationed where streets cross into the buffer zone. Also, police officers will need to be stationed at key intersections along the evacuation routes within the buffer zone to minimize delays and make traffic flow go more smoothly.

21. ***Launch*** **ArcCatalog** and go to your **student folder** in the **Catalog Tree**.

22. ***Right click*** your **student folder** and *select* **New ▸ Shapefile...** from the context menu.

    Name the new shapefile **Police_Pts** with **Point** feature type.

    Click Edit... to specify a coordinate system for the new data layer and Import... to ***navigate*** to the **C:\STARS\GIS_RS_Tools** folder and ***choose*** the spatial reference for the **Atlanta_Rds_Clip.shp** file.

    Click OK in the **Spatial Reference Properties** dialog box.

    Click OK in the **Create New Shapefile** dialog box to add the new shapefile to your student folder.

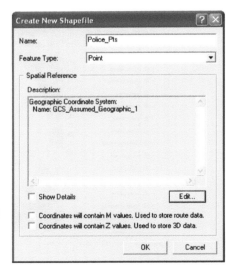

Adding a Field to the New Data Layer Table in ArcCatalog:
Because you will have two (2) different types of police stationing points in the buffer zone, you will need to add a field to the attribute table for the new data layer so that the type of point can be specified – either Direct Evacuation or Block Access. You can add a new field to the attribute table once the layer is added to ArcMap but you will add the field in ArcCatalog for this exercise. In a later lesson, you will learn to add a field to the attribute table in ArcMap.

23.   When the new **Police_Pts** layer is added to your student folder, *right click* it in the **Catalog Tree** and *select* **Properties**…

   Click the **Fields** tab. Notice the fieldnames that are automatically added to the layer's attribute table when you create a new shapefile.

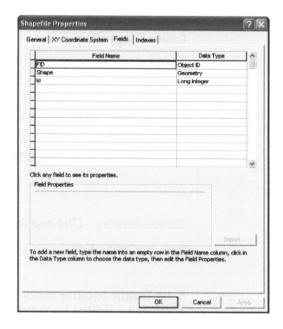

24.   To add another field to the table, *click* in the **fieldname box** under the **Id** fieldname and *type* the fieldname **Type_Statn**. In the **Data Type** box, *select* **Text**.

   Accept the **default field length** of **50**.

   Click  .

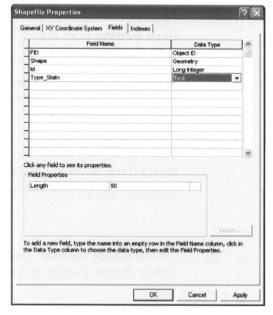

25. Make **ArcMap** active again. **Add** 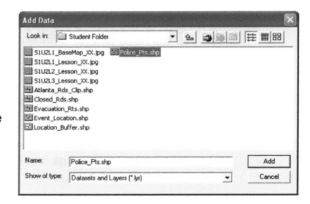 the new **Police_Pts.shp** data layer from your **student folder**.

26. When the new layer is added, **double click** the layer in the **Table of Contents** to open its **Layer Properties**.

27. **Change** the **Layer Name** to **Police Points**. In **Symbology**, **click** the **symbol**

    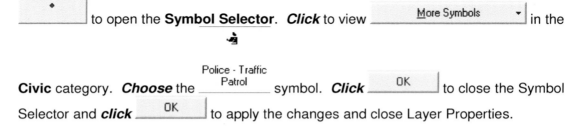 to open the **Symbol Selector**. **Click** to view More Symbols ▾ in the

    Civic category. **Choose** the Police - Traffic Patrol symbol. **Click** OK to close the Symbol Selector and **click** OK to apply the changes and close Layer Properties.

28. If the **Editor toolbar** is not currently visible in the **ArcMap** window, **select Toolbars ▸ Editor** from the **View** menu.

29. To begin an editing session, **select Start Editing** from the Editor ▾ toolbar drop-down menu.

    Select your **student folder** as the folder where the data is contained that you want to edit and **click** OK.

    When prompted that some of the data layers have differing coordinate systems, **click** Start Editing.

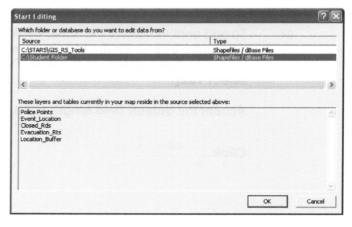

30. On the **Editor toolbar**, make sure that **Create New Feature** is set as the **Task** and **select** Police Points as the **Target**.

    Click the **Sketch** tool to heads-up digitize the first point.

    First, **click** to add a point at the intersection of the major roads in the buffer zone.

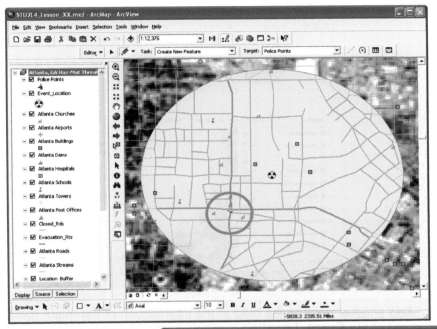

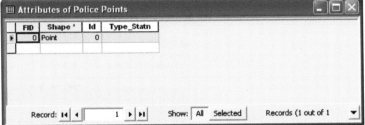

With the point you just added still selected, *right click* the **Police Points** data layer in the **Table of Contents** and *open* the layer's **attribute table**. The record for the point that you just added will also be selected in the table.

31. *Click* in the selected box under the **Type_Statn** field and *type* **Direct Evacuation**. *Press* **Enter** on your keyboard.

32. *Adjust* the size of the **attribute table window** and *move* it so that you can also see the map display.

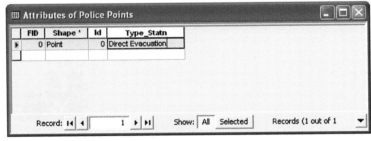

33. *Add* another point to the map display at the intersection of the major road (evacuation route) just south of the **Event_Location** symbol.

In the **attribute table**, *type* **Direct Evacuation** again (or copy and paste from the previous cell you typed).

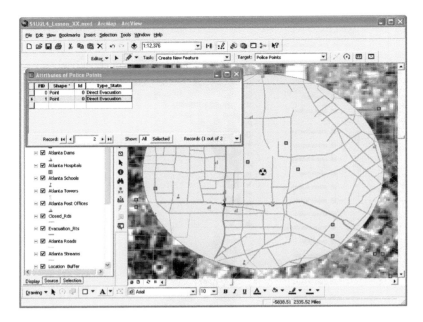

34.   ***Add*** another point to the map at the intersection of the major road (evacuation route) just west of the **Event_Location** symbol. In the **attribute table**, the ***Type*** of this point should be **Direct Evacuation** again.

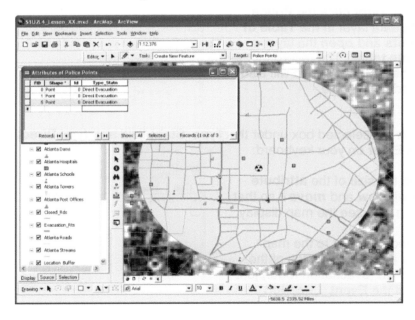

Now you will begin to add points where the boundary of the buffer and streets intersect. These police points will be added to note places where access to the area will be blocked.

35.   ***Add*** a point at one of the intersections of the **buffer boundary** and a **street**. In the attribute table, note this **type** as **Block Access**.

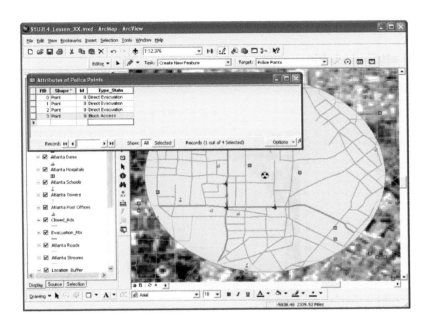

36. *Add* points at every place where the buffer boundary crosses a road segment. The **type** notation in the attribute table for each of these points should be **Block Access**.

(Note: You may find that by placing a police point at an intersection just outside the buffer zone, you may be able to use one police officer to man an area that otherwise would require two offices, thus reducing the number of police officers handling traffic flow and making more available at the crisis scene. See the example that follows.)

You may need to use the **Zoom In** and **Pan** tools to view the area more closely.

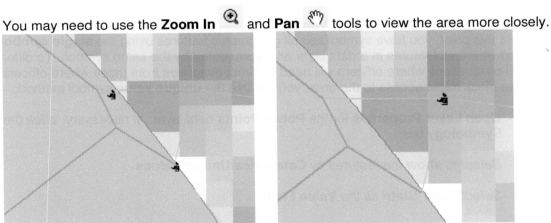

By moving the access point just outside the buffer zone, personnel resources needed for traffic control can be reduced.

If, at any time, you need to delete a feature that you added, *select* the **Select Features** tool , *click* the feature and *press* the **Delete** key on your keyboard.

When you finish, you should have approximately 40 points in this data layer (depending on where you placed your points).

Close the **attribute table**.

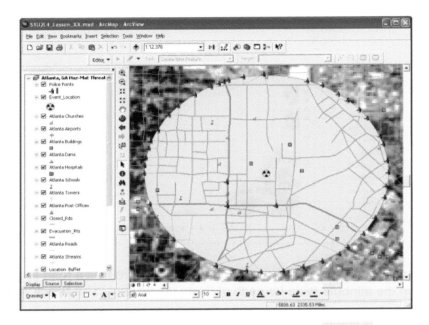

37. To stop the editing session, ***select*** **Stop Editing** from the Editor ▼ toolbar drop-down menu and ***click*** Yes to save your edits.

Symbolizing Features by Unique Value:
Up until this point, you have symbolized all geographic features using the **single symbol** method where all features in a data layer are represented by the same symbol. To differentiate the police locations where officers will need to direct evacuees from those where officers will need to block traffic access to the area, you will use the **unique value** symbol method.

38. ***Open*** **Layer Properties** for the **Police Points** data layer. If necessary, ***click*** the **Symbology** tab.

Select to **show** the features by **Categories/Unique values**.

Select **Type_Statn** as the **Value Field**.

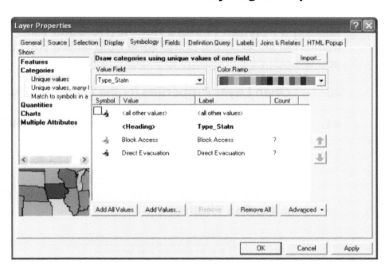

Click to Add All Values . Notice that when you do this, the symbols revert back to plain points instead of police officer symbols. To change this, *click* on the Symbol header and *select* **Properties for All Symbols…** *Choose* the

Police - Traffic Patrol symbol again and *click* OK to close the **Symbol Selector**.

In **Layer Properties**, *change* the **color** of the **Block Access symbol** to Red. *Change* the **color** of the **Direct Evacuation symbol** to **Black**.

Deselect the **checkbox** □ to display <all other values> because there are no other values in our table.

Click OK to apply the changes and *close* Layer Properties.

39.   Click 💾 to save your project.

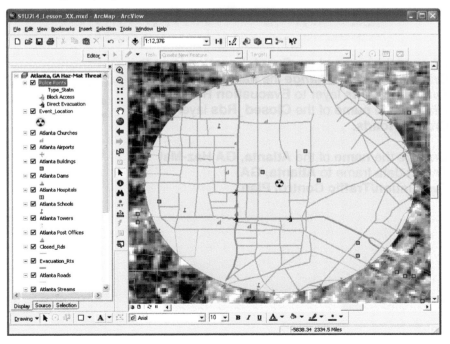

Making a Polygon Feature Hollow (No Color Fill):
Instead of showing the buffer zone filled where you cannot see anything inside the buffer area, you will edit its symbology to make it hollow.

40.   *Open* **Layer Properties** for the **Location_Buffer** layer.

41.   *Change* the **name** of the layer to **Half Mile Buffer Zone**.

Click the **Symbology** tab and *click* the ⬜ symbol to open the **Symbol Selector**. *Change* the **Fill Color** to **Hollow**. *Change* the **Outline Color** to **Yellow** with an **Outline Width** of **4**.

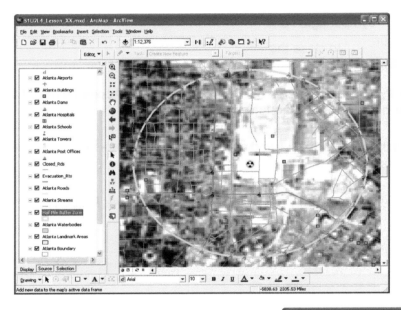

42.   *Change* the **name** of the **Event_Location** layer to **Crisis Location**. *Change* the **name** of the **Evacuation_Rts** layer to **Evacuation Routes**. *Change* the **name** of the **Closed_Rds** layer to **Closed Roads**.

43.   *Change* the **name** of the **Atlanta, GA Haz-Mat Threat** data frame to **Atlanta, GA Evacuation/Traffic Control Plan**.

Creating the Map Layout:

44.   *Switch* to **Layout View**.

45.   *Edit* the title at the top of the layout page to have a subtitle to the last title so that it reads:

Atlanta, GA Haz-Mat Threat
Evacuation/Traffic Control Plan

46.   *Double click* the **legend** to open **Legend Properties**.

    Click the **Police Points** layer in the **Legend Items** list and *click* the **Promote** button repeatedly to move this point data layer up to the top of the list if needed.

    Adjust the order of any other legend items.

    When you are finished, *click*   OK .

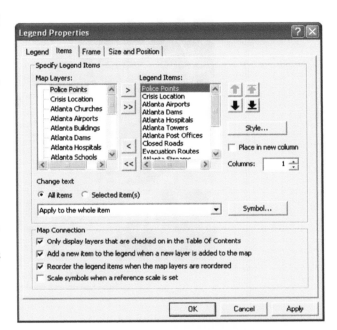

47.   *Delete* and *reinsert* the **scale bar** for this map layout.

48.   *Edit* the date in the text box containing your name and date.

49.   *Export* the layout to **your student folder** in JPEG format as **S1U2L4_Lesson_XX** (where XX is your initials).

50.   *Print* the map.

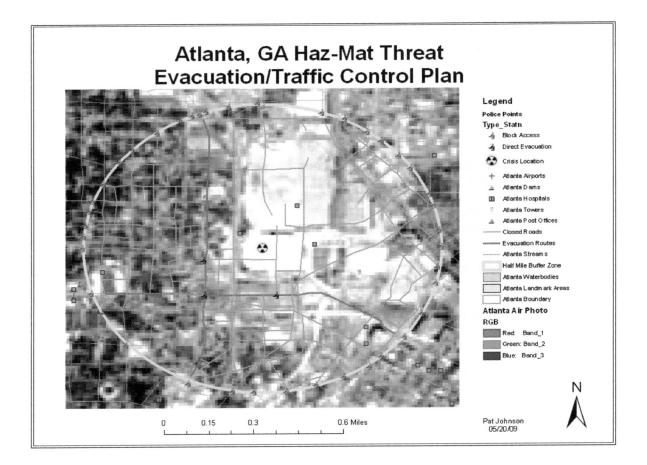

Saving & Closing:

51.   ***Save*** this project.

52.   ***Select* Exit** from the **File** menu or ***click*** the **Close** ☒ button in the upper right corner of the ArcMap window to exit.

Lesson 4: Analyzing Geospatial Data Lesson Review

Key Terms
Use the Lesson and the Lesson Activity to define each of the following terms.

1.   **ArcToolbox -**

2.   **Buffer** -

3.   **Identify tool** –

4.   **Sketch tool** –

Global Concepts
Use the information from the lesson to answer the following questions. Use complete sentences for your answers.

5.   Name at least two instances in which buffers could be used.

6.   What is the difference between using a single symbol and using the unique value symbol method?

Let's Talk About It...
Answer the following question and share the responses with your instructor and classmates.

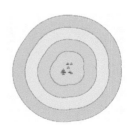

7.   Another type of buffer that can be used is called a Multiple Ring Buffer which when applied looks similar to a target. What do you think could be a reason for using a multiple ring buffer in a map display? Be prepared to explain your answer.

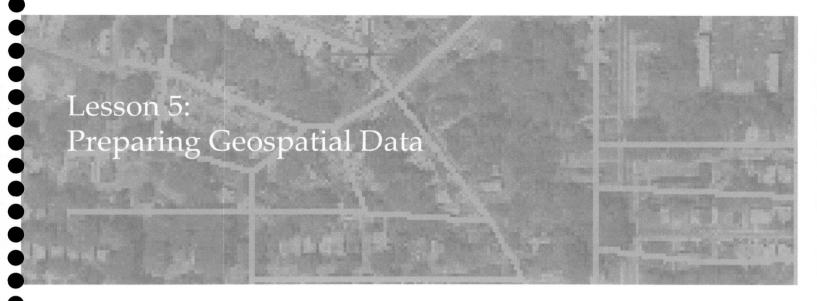

Lesson 5:
Preparing Geospatial Data

With geospatial software like ArcMap, you can create many different maps for many different situations. Whether you create the data you need for your study, find it on the Internet, or buy it from another source, you must consider the spatial reference of the data. So what is spatial reference and why is it important? Simply put, spatial reference refers to a feature's geographic location on the Earth using map projections and coordinate systems. Anytime you show data in a basic GIS, you are taking three dimensional information and displaying it on a two dimensional surface. In order for data layers to align properly in a GIS program, the data should all use the same coordinate system. The coordinate system that people are most aware of is the Geographic Coordinate System where location is given in degrees of longitude and latitude. However, as we have grown to know more and more about our world and its shape, scientists have developed different models of the Earth and different ways of communicating location have grown out of these efforts.

Reproject "On the Fly"

ArcGIS provides a reproject "on the fly" capability that will allow data with differing spatial references to be displayed together. For example, if you have a data frame that is projected in a UTM projection and add a data layer that has a geographic coordinate system - like a WGS 1984 shapefile - the shapefile will be reprojected on the fly. It is important to note that ArcMap will reproject the data for the sake of display purposes only. The spatial reference of the original source data file is not changed at all. The user does have the capability to edit the data if needed.

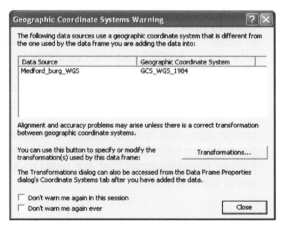

An example of the warning you will receive when the data layer being added has a different geographic coordinate system the data in the project.

Layers without Spatial Reference

There will be times when you find data layers that have missing spatial reference. These layers must be given a spatial reference before they can be used in a GIS project. Also, image data that is collected (from air photographs or satellite images) must be given spatial reference before they can be used.

Georeferencing

Spatial reference can be assigned to image data through a process called georeferencing. By aligning an image to another data layer that has a spatial reference, you give the image its own spatial reference and ensure that the image layer will align properly with other data layers of the same geographic area.

Georeferencing is performed in ArcMap. There is a separate Georeferencing toolbar that contains all of the tools necessary to georeference a non-georeferenced image layer.

This image is being spatially referenced to a street file using the georeferencing tool in ArcMap.

Creating a New Shapefile

In this lesson, you will create a new shapefile in ArcCatalog. Instead of creating a point shapefile as you did in an earlier lesson, you will create a polygon shapefile. Once the new shapefile is added to your map layout, you will start an editing session and add polygon features to the new data layer using "heads-up" digitizing techniques in ArcMap. Because it is important to be able to see the area beneath the polygons that you will create, you will adjust the transparency of the polygons using the Transparency tool.

Data Inventory

In this particular lesson, you will explore georeferencing methods as you georeference an image layer to a vector street layer. You will also create a new shapefile and edit its transparency. All of this will be explored using techniques centering on a Homeland Security scenario. All of these techniques and skills although they are being introduced using a Homeland Security scenario, can also easily be used for geospatial analysis in other disciplines as well.

Conclusion

In order for data to align properly in a map display, the layers must use the same spatial reference. ArcGIS has the capability to reproject data "on the fly" to allow you to layer data with differing spatial references in the same map display. The spatial reference for the layers that are reprojected on the fly is not changed although they can be edited by the user. The reproject "on the fly" function will not work on data that has no spatial reference.

Image data that has no spatial reference cannot be properly displayed unless it is georeferenced to a data layer that has the necessary spatial reference. The Georeferencing toolbar contains tools that are used to align the data layer to an already georeferenced layer using control points.

Lesson 5: Preparing Data

In the last four (4) lessons, you may have noticed that one of the data layers that is included in the project is an aerial photograph of Atlanta, GA. Because the image includes a relatively large geographic area, the resolution of the image is only 12 meters – meaning that one (1) pixel in the image represents 12 meters on the ground. The photo looks fine when you are zoomed out to the full geographic extent of the study area. However, when you zoom in on the smaller area that you have been studying, the image becomes pixilated and it is difficult to interpret features in the image. For this reason, you will add another image to the project that has a higher resolution and will be better to use for the smaller geographic area that has become the focus.

The one problem that exists in using this image is that it is not **georeferenced**. In other words, the image does not have a spatial reference associated with it. When an air photo is taken, it is simply a digital image just like any other photo. Only when you align the image to a data layer (image or feature layer) that has a spatial reference associated with it can it be used in a GIS.

In ArcGIS 9.3 when data has a defined coordinate system (spatial reference) and is added to a project that has other data layers that have a different defined coordinate system, ArcGIS will reproject the added data layer "**on the fly**," meaning that the data will be manipulated so that it can be displayed with the other data. For example, a data layer that uses the Universal Transverse Mercator (UTM) coordinate system will align in the map display with layers that use the Geographic Coordinate System. Keep in mind that the source data file for the UTM layer as well as the source coordinate system for it are not changed at all – only the way the data is displayed is changed for the purpose of the project.

A data layer that has no spatial reference at all must be georeferenced or have a spatial reference assigned to it in order for it to display with other layers properly. You will gain some experience with these concepts and with the skill of georeferencing in this lesson.

Although you will use the same project that you have worked with in earlier lessons in this unit, you will georeference the air photo in a new, separate document. This process could be done in the map document that you worked with earlier; however, the number of layers that was included in the project could make the process more cumbersome. Once the image is georeferenced, it will be added to the earlier project.

Exploring Spatial Reference:
You will begin by comparing spatial references for some of the different data layers that you will be using in this lesson.

1.   *Launch* **ArcCatalog**.

2.   In the **Catalog Tree**, *open* the **C:\STARS\ GIS_RS_Tools** folder and *click* on the **Atlanta_Rds_Clip.shp** data file.

In the **Preview** window, *click* on the **Metadata** tab.

Click the [Spatial] sub-tab to display the **spatial metadata** for this data layer.

Terminology Tip

Metadata provides descriptive information about geographic data including information about data quality, data organization, spatial reference, entity and attribute information and contact information.

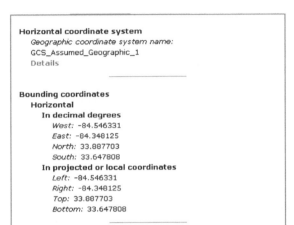

The coordinate system for this particular data layer is GCS (Geographic Coordinate System or longitude/latitude). In the Bounding Coordinates section, you see that the coordinates are given in decimal degrees.

3. In the **Catalog Tree**, *click* on the **Atlanta_Rds_UTM.shp** layer. *Click* to view this layer's [Spatial] metadata.

Notice that this layer (that contains the same features as the **Atlanta_Rds_Clip.shp** data layer) is projected in UTM with the 1983 North American datum.

Horizontal coordinate system

Projected coordinate system name:
NAD_1983_UTM_Zone_17N
Geographic coordinate system name:
GCS_North_American_1983

Horizontal coordinate system
Projected coordinate system name:
NAD_1983_UTM_Zone_17N
Geographic coordinate system name:
GCS_North_American_1983
Details

Bounding coordinates
Horizontal
In decimal degrees
West: -84.553861
East: -84.339145
North: 33.888977
South: 33.642504
In projected or local coordinates
Left: 171279.814547
Right: 190303.715323
Top: 3754895.570213
Bottom: 3728155.213305

The coordinates for the source data file were in decimal degrees.

In decimal degrees
West: -84.553861
East: -84.339145
North: 33.888977
South: 33.642504

The projected coordinates are provided and, for UTM, these coordinates are given in the unit of measure of meters.

In projected or local coordinates
Left: 171279.814547
Right: 190303.715323
Top: 3754895.570213
Bottom: 3728155.213305

4.   In the **Catalog Tree**, *click* on the **Atlanta_GADome_2m.jpg** data file. This is the non-georeferenced image that you will georeference later in this lesson.

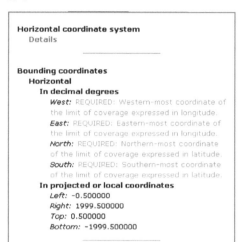

Click to view this layer's **Spatial** metadata.

Notice that it has none. Arbitrary coordinates are given to the file based on file size and other factors. No information about coordinate systems or bounding coordinates in decimal degrees can be given because it does not have spatial reference – it is just a photo at this point.

Georeferencing an Image Data Layer:
When you georeference a data layer, you can use either an image layer or a feature layer as the reference layer. Put simply, you find corresponding points in each data layer and align them. For instance, if you see a major intersection on an image, you can find that intersection in the other reference data layer and align the points. You continue to find "control points" until the data layer is properly aligned.

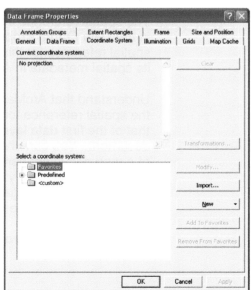

5.   *Click* the **Launch ArcMap** button to open ArcMap and begin the georeferencing of the air photo.

6.   This time in the **ArcMap** dialog box, *click* to start using ArcMap with **⊙ A new empty map** and *click* **OK**.

7.   When the blank map document opens, *double click* the **Layers** data frame to open **Data Frame Properties**. *Click* the **Coordinate System** tab.

Notice that no coordinate system is listed. This is to point out that ArcMap does not automatically assign a coordinate system or have a "default" coordinate system that it uses for all projects.

***Close* Data Frame Properties**.

8.   ***Add*** 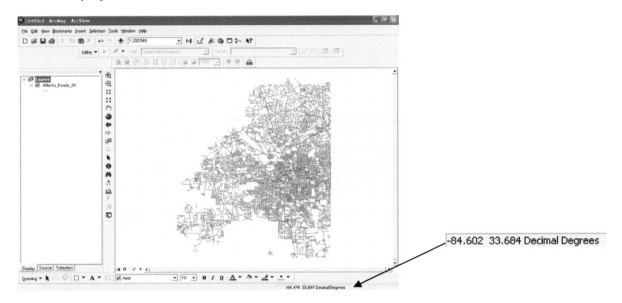 the **Atlanta_Rds_84.shp** data layer from the **C:\STARS\GIS_RS_Tools** folder.

Notice at the bottom of the map display that the units for the data frame are displayed in Decimal Degrees and that the values change as you move your mouse across the map display.

-84.602 33.684 Decimal Degrees

9.   ***Open* Data Frame Properties** again and, if necessary, ***click*** the **Coordinate System** tab.

The spatial reference information is now displayed in the Data Frame Properties, just as you saw the spatial reference for this layer when you viewed its spatial metadata in ArcCatalog.

Understand that ArcMap automatically assigns the spatial reference for the map document as that of the first data layer added to a map document. You can change the spatial reference of a map document here in the Data Frame Properties if necessary. However, you will georeference the image to WGS 84 so the coordinate system does not need to be changed.

***Close* Data Frame Properties**.

Because the image that you will be using does not include the entire city of Atlanta, you will select and then export a specific area of the Atlanta street network data layer that includes the geographic area that is contained in the image.

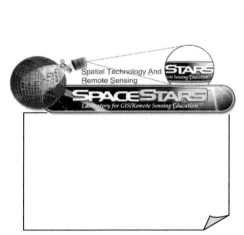

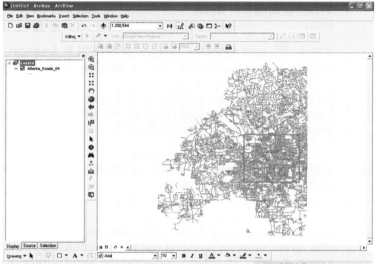

This is a form of **clipping** and is a process that makes georeferencing an image to the feature layer easier because you do not have as many road segments to choose from when finding control points to align. A more precise form of clipping, using ArcToolbox, will be covered in a later lesson.

The image you will be using is of the Georgia Dome area. This area is identified in the following map display.

10.   Using the **Select Features** tool , *click* and *drag* this area with your mouse so that the road segments in the area are highlighted.

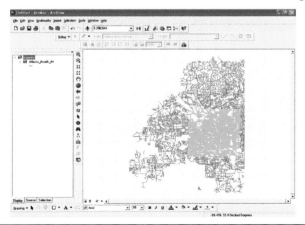

Tool Tidbit

Select Features - This toolbar button allows you to interactively select features in the map display.

You will now export these selected features as a new data layer.

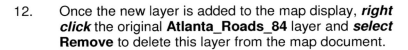

11. **Right click** the **Atlanta_Roads_84** data layer and **select** **Data ▶ Export Data...** from the context menu. **Save** the new data layer as **GA_Dome_Clip.shp** in your student folder.

    When prompted, **click** Yes to add this data to the map.

    Click 💾 and **save** the project as **S1U2L5_Lesson_XX** (where XX is your initials) in your **student folder**. Keep the project open.

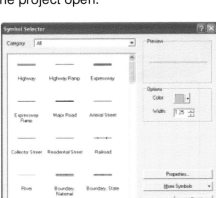

12. Once the new layer is added to the map display, **right click** the original **Atlanta_Roads_84** layer and **select** **Remove** to delete this layer from the map document.

13. **Click** the **Full Extent** button ⬤ so that the new street network clip fills the display.

14. Once the image is added to the display, you will need this street network to stand out to make finding control points easier. To do this, **change** the **symbology** of the **GA_Dome_Clip** layer to an Orange line with **line width** of **1.25**.

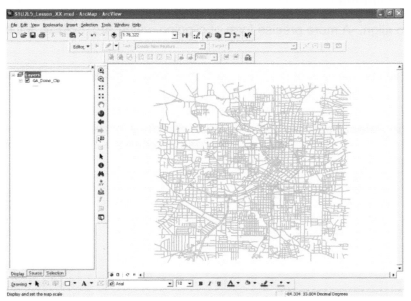

---

15. ***Add*** 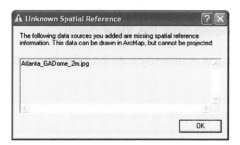 the **Atlanta_GADome_2m.jpg** data layer from the **C:\STARS\GIS_RS_Tools** folder. The following warning dialog box will appear:

Click OK . The image data layer will be added to the map document but will not display in the map display with the street network clip layer that is displayed. When you think about it, this makes sense. The street network clip is displayed in Decimal Degrees somewhere in the range of -84.447129 in the left, -84.254425 dd right, the top is at 33.800405 and the bottom is at 33.709107 of the layer. The image layer was given arbitrary coordinates of –0.5, 1999.5 and 0.5, -1999.5 in unknown units for its bounding coordinates.

To illustrate this point, ***right click*** the **Atlanta_GADome_2m.jpg** data layer and ***select* Zoom to Layer**.

If you place your mouse in the lower left corner of the image, you will read 2.936 negative 1972 and in the upper right corner you will read 2004.021 -2.941. In other words, ArcMap takes the coordinates and attaches Decimal Degrees to those coordinates because, since the first data layer added was in Decimal Degree units, it attaches those units to this layer because it has no spatial reference. It is important to note that both layers are displayed. However, because their bounding coordinates are so different, they are nowhere close to aligning.

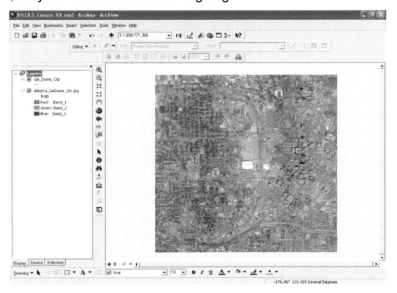

---

16.   Because the street network is our reference data layer for the georeferencing process, *right click* the **GA_Dome_Clip** layer and *select* **Zoom to Layer**.

17.   In order to georeference the image layer to this street network, you will need to use the tools contained on the **Georeferencing toolbar**. To access this toolbar, *select* **Toolbars ▸ Georeferencing** from the **View** menu.

       Notice that the image layer is already listed as the layer to be georeferenced.

18.   To begin the georeferencing process, *select* **Fit to Display** from the Georeferencing ▾ toolbar drop-down menu. ArcMap will attempt to roughly align the image to the street network.

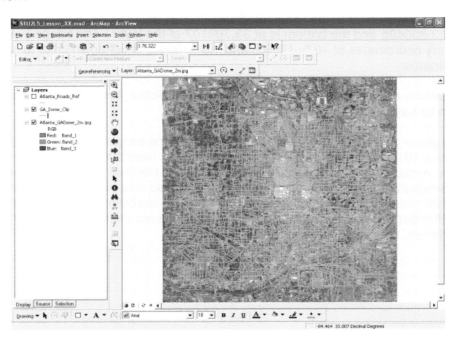

       You will now have to select control points in the image that you can use to adjust the image to the street network. Keep in mind that the street network is currently in the projection that we want to use for the image. So you will always **click the control point on the image first** (the From point) and **then click on the corresponding control point on the street network** (the To point). In other words, you will be telling ArcMap to move the image from this first point to match it up with the point on the street file.

       You will note that there are several noticeable characteristics in the street network – major intersections, major highways and interstates, curves in roads, etc. You will use these as your control points because they will be easy to distinguish in both the image and the street network.

Introduction to GIS Tools and Processes, version 7

In ArcMap, the general rule is to use at least three (3) control points for a georeferencing iteration. Your control points should be dispersed throughout the image and not all together, as the image would align closely around the control points but distortion would be great in other parts of the image.

Because you may not be familiar with the Atlanta area, an even smaller clip of this street network is provided for you in your data folder. You will need to add this data layer and use it to georeference the image.

19. ***Click*** ✛ to add **Atlanta_Roads_Ref.shp** to the map display. ***Change*** the **symbology** of the **Atlanta_Roads_Ref** layer to a bright **Yellow** line with **line width** of **1.25**.

20. ***Right click*** and ***remove*** the **GA_Dome_Clip** from the **Table of Contents**.

For this exercise, the first control point is identified for you will be in the left side of the image.

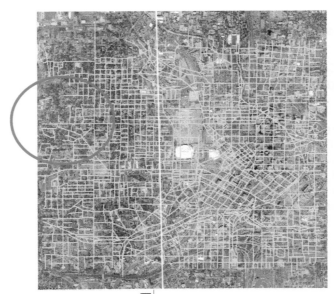

You may find it easier to ***turn off*** ☐ (uncheck) the **Atlanta_Rds_84** street network layer in the **Table of Contents** so that you can see unique street landmarks in the image easier. Keep in mind that you can use the **Identify** tool ⓘ to determine the names of streets in the street network or you can use **Select by Attributes** to actually select the interstates in the street network making it easier to identify them. In the following graphic Interstate 20 highlighted.

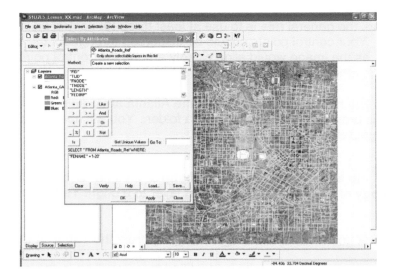

21.   **Zoom** in  closer to the area near the first control point.

22.   **Click** the **Add Control Points** tool on the
       Georeferencing toolbar.

       Focus on the street that curves as it runs from
       north to south. Find the "elbow" on that street in
       the image. This is the place where you will place
       your first control point. (You may want to turn off
       your street layer to see this point. After selecting it
       on the image, turn on the layer in the Table of
       Contents.)

       Click the street in the image as shown in this
       image.

If you turned off the street layer, turn it back on at this point. Second, **click** the corresponding point on the street network that is located to the right and slightly down in the image. As you move your mouse to the control point in the street network, you will notice a rubber band effect from the first point.

At any point after setting the From point for the first control point, you want to delete the first point, click the **Esc** key on your keyboard. If after setting the first control point, you feel that you have made a mistake, click the Delete key on your keyboard to delete the most recent control point that you have established.

ArcMap will adjust the image so that the points that you identified as the first control point align.

23.   Zoom to **full extent**  so that the second control point can be identified.

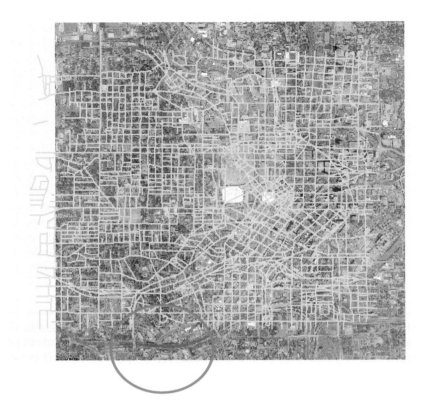

24. ***Zoom*** in to this area.

Just west of the place where two roads curve, there is an intersection that corresponds with the intersection in the image.

25. **Click** the **Add Control Points** tool and **click** on the intersection in the image.

Click the corresponding point in the street network.

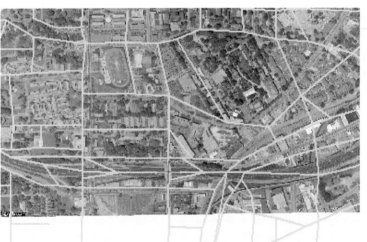

Tool Tidbit

Add Control Points ✗ -
This toolbar button allows you to set control points when georeferencing a data layer.

26. **Zoom to full extent** ⬤ so that the third control point can be identified.

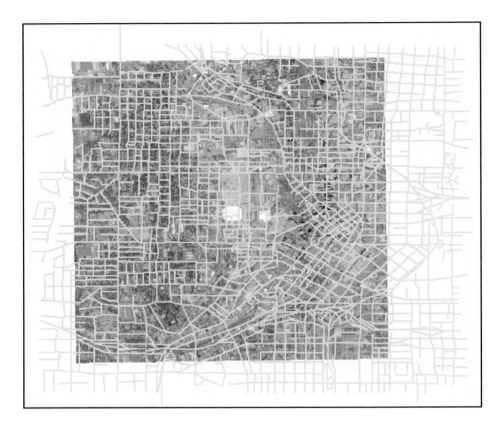

27.   In the upper right part of the image, there are several roads that curve noticeably. *Zoom* in to that area. Notice the intersection of the easternmost curved road.

28.   Use the *Add Control Points* tool and *click* the intersection in the image.

Click the corresponding point on the street network.

Notice that the image is more closely aligning with the street network now.

29.   ***Click*** 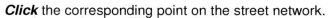 to ***Zoom to full extent***.

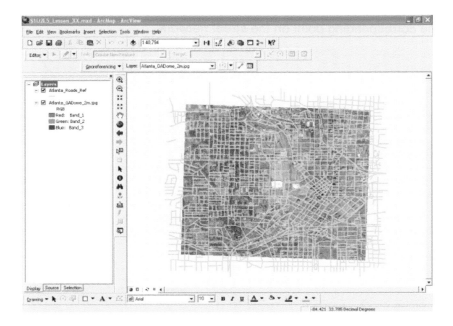

30. ***Turn off*** the **street network** in the **Table of Contents** so that you can view the image.

Notice the placement of the first three (3) control points. You will need to add the fourth control point in an area that doesn't already contain one. A good place would be in the southeastern (bottom right) portion of the image.

Zoom to this area.

31.   Turn on the **street network** layer. The fourth control point will be located at an intersection just east of the baseball field.

The image will even more closely align. However you might experience a small blue line between the two points.

When establishing the accuracy of the control points, a mathematical measurement can be calculated comparing the actual location of the map coordinate to the transformed position in the image. The distance between the control points is called the **residual error**. This point will have a higher residual error than others which can be seen by the blue mark between the two points.

Total error is calculated by summing all residuals and taking the root mean square (RMS) of the residual values.

As you add more control points, the image has to adjust while attempting to retain the points that you have already added. This may cause some stretching of the original points as more points are added.

32. To access this information, **click** the **View Link Table** button 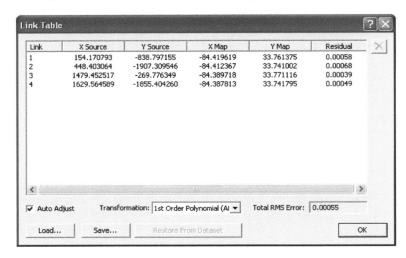 on the **Georeferencing** toolbar. The **Link Table** dialog box will appear.

The Total RMS error in this example is relatively low at 0.00055 however it could be better by being less. This total RMS is stated in terms of the data frame's measurement which is currently in Decimal Degrees. When converted to meters, the RMS is off by over 55 meters. If your study was to concentrate on a very large area, this might not be an issue, however if your study involved being more accurate on a large scale area (i.e. mapping locations of fire hydrants), this georeference might be a problem.

If there is a problem with any of your links, you can delete the link from this list and redo the control point in the map display. To do this, highlight the link in the list and **press** the

Delete key on your keyboard or **click** the **Delete** button ☒ on the **Link Table** dialog box to delete this link. Because the last link presented a problem, you will need to delete that last link set.

Click ⬚ OK ⬚ to close the **Link Table** dialog box.

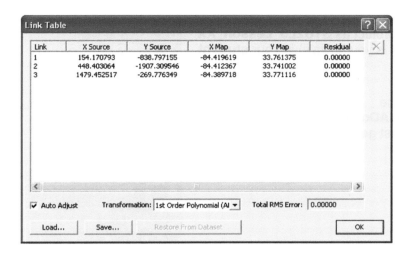

33. To save the rectified image, **select Rectify** from the 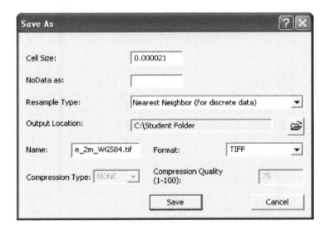 drop-down menu.

   Accept the default **cell size** and **Nearest Neighbor Resample Type** but **change** the **output raster** so that the image is saved in your **student folder** as **Atlanta_GADome_2m_WGS84.tif**.

   Verify that your settings match those above and **click** [Save].

34. **Save** this **ArcMap** map document as **S1U2L5_GeoRef.mxd** in your **student folder**.

   Now that you have georeferenced this image, you will use it in the map that you created for Lesson 4.

35. **Click** the **Open** button on the standard toolbar to open the **S1U2L4_Lesson_XX.mxd** map document that you worked with in the last lesson.

36. **Switch** to **Data View**.

Save the project as **S1U2L5_Lesson_XX** (where XX is your initials) in your **student folder**.

Add the **Atlanta_GADome_2m_WGS84.tif** image that you just georeferenced to this project.

37.   **Zoom** to the **Atlanta_GADome_2m_WGS84.tif** layer.

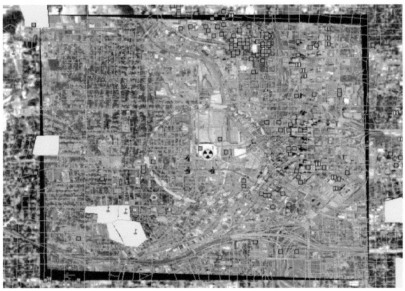

Knowledge Knugget
For all images, the color of each pixel in the image is a blend of three colors– red, green and blue – when using RGB color scheme. By blending these three colors, many different colors can be created.

38.   **Open Layer Properties** for **Atlanta_GADome_2m_WGS84**. **Click** the **Symbology** tab.

Check to ☑ Display Background Value:(R, G, as **No Color**.

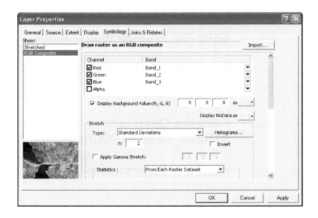

Click ___OK___ .

The border that was displayed around the **Atlanta_GADome_2m_WGS84.tif** image will become transparent.

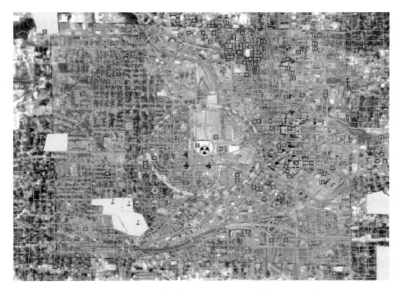

39.   ***Zoom*** back in on the crisis area. Notice how much better the resolution of this image is compared with the 12-meter image of the entire city of Atlanta.

As you can see from the high-resolution air photo that you added to the map document, there are some larger, open areas around the event site that could be used as staging areas.

Some of these areas, particularly parking lots near the stadium, would be ideal staging areas once event attendees were evacuated. In order to map these areas, a new data layer must be created in ArcCatalog.

40.   *Click* the **Launch ArcCatalog** button .

In the **Catalog Tree**, *create* a **new shapefile** in your **student folder**. This time, **name** the new layer **Staging_Areas** with **Polygon feature type**.

Specify the same **spatial reference** as the **Atlanta_Rds_Clip.shp** layer in the **C:\STARS\GIS_RS_Tools** folder.

41.   *Return* to **ArcMap** and *add* 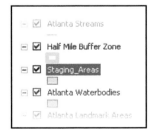 the **shapefile** that you just created from your **student folder**.

In the **Table of Contents**, *click and drag* the **Staging_Areas** layer below the **Half Mile Buffer Zone** layer.

42.   To add data to this new data layer, you must begin an editing session. You have added features to new layers in previous lessons. However, until this point you have "heads-up digitized" point features. You will now digitized polygon features.

Select **Start Editing** from the **Editor** ▼ toolbar drop-down menu. *Select* your **student folder** from the **Start Editing** dialog box and *click* OK .

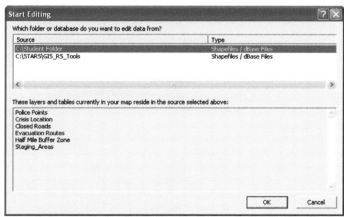

When prompted about differing coordinate systems among the data layers in the project, *click* Start Editing .

43. In the Editor toolbar, make sure that the **task** is set as **Create New Feature**. The **target** layer should be **Staging_Areas**.

Zoom in on the **lawn area** just east of the Georgia Dome.

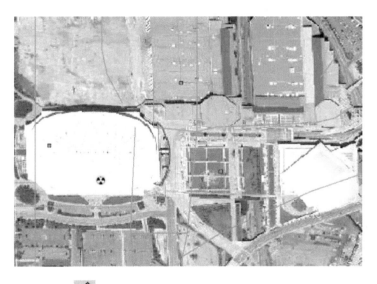

44. *Click* the **Sketch** tool ✏, take your mouse and *click* around the **boundary** of the grassy area to create the polygon for this new layer to mark the area where responders could prepare and assemble as necessary during the crisis. When you make the **last click**, *double click* your mouse.

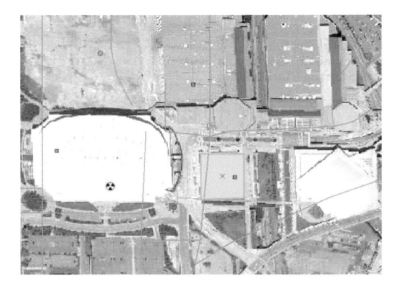

45.   ***Create*** another **polygon** in the triangular-shaped area just south of the area you just
        digitized.

---

46. *Continue digitizing* open areas inside the buffer zone that could be used as **staging areas**. (Keep in mind that you may digitize the construction area just north of the Georgia Dome, but buildings have been located there since this air photo was taken).

When you finish, you should have several open areas near the Georgia Dome digitized.

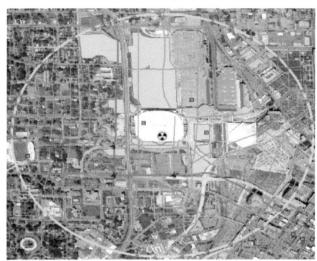

47. When you have finished, *select* **Stop Editing** from the  toolbar drop-down menu and *save* your **edits**.

Sometimes when you are dealing with feature data layers and image data layers in the same project, it helps to alter the transparency of one or more features layers so that you can see the imagery underneath. You will change the transparency of the **Staging Areas** data layer so that you can see the air photo below the layer.

48. To display the **Effects toolbar**, *select* **Toolbars ▸ Effects** from the **View** menu.

49.   In the **Layer box**, *select* **Staging_Areas**.

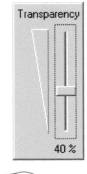

Click the **Adjust Transparency button** 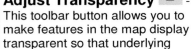 and *adjust* the **slider bar** to **40%**.
The **Staging_Areas** polygons will become transparent in the map display.

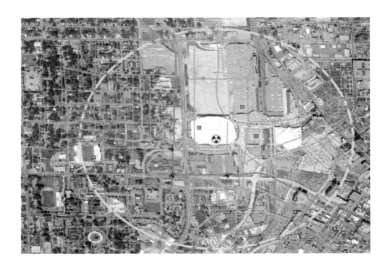

Tool Tidbit

Adjust Transparency 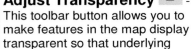 -
This toolbar button allows you to
make features in the map display
transparent so that underlying
layers can be seen.

50.   *Change* the **layer name** of the **Staging_Areas** data layer to remove the underscore
from the name and change the name of the **Atlanta_GADome_2m_GCS.tif** to **Georgia
Dome Air Photo**.

Creating the Map Layout:
51.   *Switch* to **Layout View**.

52.   *Edit* the title at the top of the layout page to have a subtitle to the last title so that it
reads:

Atlanta, GA Haz-Mat Threat
Response Staging Areas

53.   In the **Table of Contents**, *turn off* ☐ the **Atlanta Air Photo**, because it cannot be seen
in the data frame when zoomed in to the Georgia Dome.

54.   *Delete* and *reinsert* the **scale bar** for this map layout. .

55.   *Edit* the date in the text box containing your name and date.

56.   *Export* the layout to **your student folder** in **JPEG** format as **S1U2L5_Lesson_XX**
(where XX is your initials).

57.   *Print* the map.

Saving & Closing:

58.   *Save* this project.

59.   *Select* **Exit** from the **File** menu or *click* the **Close** ✖ button in the upper right corner of
the ArcMap window to exit.

Lesson 5: Preparing Data Lesson Review

Key Terms
Use the Lesson and the Lesson Activity to define each of the following terms.

1.   **Georeferencing –**

2.   **Metadata –**

3.   **Select Features –**

4.   **Clipping –**

5.   **Residual Error –**

6.   **Total Error –**

Global Concepts
Use the information from the lesson to answer the following questions. Use complete sentences for your answers.

7.   Why would the "on the fly" feature be useful in displaying data? Does it permanently alter data?

8.   Explain the procedure to set a control point in georeferencing an image to a street file. How many control points should you set?

Let's Talk About It...
Answer the following question and share the responses with your instructor and classmates.

9.   Describe what happens if too many control points are used or they are placed too close together. What can be done to correct the situation if this happens?

Lesson 6:
Planning & Building a Local Data Inventory

In the last five (5) lessons, you have been introduced to many tools, techniques and skills pertaining to the ArcGIS 9.3 software suite. In each of those lessons and with each of the activities contained in those lessons, you used demonstration project data that focused on the Atlanta, GA area. By this point, you probably have a good idea of the tremendous benefits that can be realized by having and organizing a data inventory of your community. Building a data inventory is much like creating a digital community. A data inventory provides a snapshot of your community in a "normal" state or the "base map," while giving you the opportunity to simulate "abnormal" circumstances (like you did with the Homeland Security scenarios in the previous lessons) to be better prepared in case crises arise.

In this lesson, you will focus on your community. You will start building a data inventory of your community. After adding, downloading, managing and preparing the data that you will need to begin building the data inventory for your community, you will have the opportunity to practice some of the skills you have learned in the previous lessons of this unit using your local data. Keep in mind that you won't build an entire data inventory in this lesson. As you continue to learn additional geospatial analysis skills in other units using other industry focus areas, you will continue to add to this inventory and then experiment with the data using the tools and techniques contained in the demonstration lessons.

Building a Data Inventory
In this activity, you will add data that was delivered with your courseware as well as download and create data of your own. Keep in mind that the image data layer that you will use is georeferenced, meaning that the image layer has a spatial reference associated with it. You will not have to georeference the layer yourself, as you did in the previous lesson with the image of the Georgia Dome in Atlanta.

Layering Existing Local Data:
You are just now beginning to build the data inventory for your community. Therefore, you will start this ArcMap project with a new empty document and add the necessary layers.

1.   *Launch* **ArcMap** and start with a **new empty map**.

2.   *Add*  the **XXXX_YYY.tif** (where **XXXX** is your county abbreviation and **YYY** is your school abbreviation) air photo and the **XXXX_321.tif** satellite image that are located in the **C:\STARS\GIS_RS_Tools\Local** folder. When you navigate to the **Local** folder, *hold* the **Ctrl** key on the keyboard and *select* both image files to add them both at the same time.

If necessary, *click and drag* the **satellite image** of the county below the **air photo** of your school in the **Table of Contents**.

NOTE: The images following are for reference only and are not representative of your specific area.

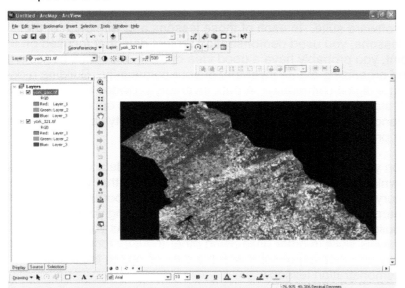

Knowledge Knugget

The **XXXX_321.tif** satellite image that you added to this project is a composite Landsat image made up of three bands of data – bands 1, 2 and 3. Composite images made up of bands 1, 2 and 3 are considered "real color" images because they display data as it actually appears.

(If a black border appears surrounding the satellite image, *open* **Layer Properties** for the satellite image and, under **Symbology**, *click*

to ☑ Display Background Value:(R, G, B)   0   0   0   as ▢ ▾   as **no color**.)

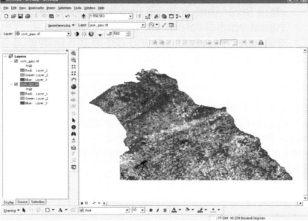

---

3. ***Rename*** the air photo image layer **YourCity Air Photo** and the satellite image layer **YourCounty Landsat Image**.

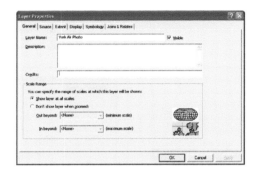 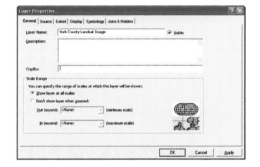

4. ***Add***  the **XXXX_tgr#####plc00.shp** (where **XXXX** is your county abbreviation and **#####** is the TIGER code for your county) data layer that is located in the **C:\STARS\GIS_RS_Tools\Local** folder.

This data layer contains all of the city boundary polygons in your county.

Knowledge Knugget

The Census Bureau collects feature data that is known as TIGER (Topologically Integrated Geographic Encoding and Referencing) data. Each state has its own numeric code and each county has its own numeric code that makes up the 5-digit code that all TIGER data filenames include.

5.　Use the **Select by Attributes** option from the **Selection** menu to *create a query* to select **your city** from the data layer.

　　Select **Zoom to Selected Features** from the **Selection** menu.

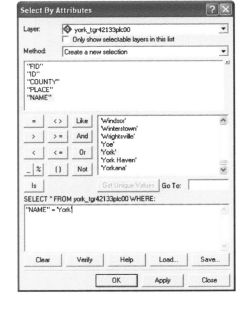

6.   ***Export*** the **selected feature** as a new shapefile. ***Name*** the new shapefile **YourCity.shp** and save it in your **student folder**.

When prompted, ***click*** [Yes] to add the shapefile as a layer to the map.

Right click the **XXXX_tgr#####plc00.shp** data layer containing all of the city polygons and ***select*** **Remove** to delete this layer from the map display.

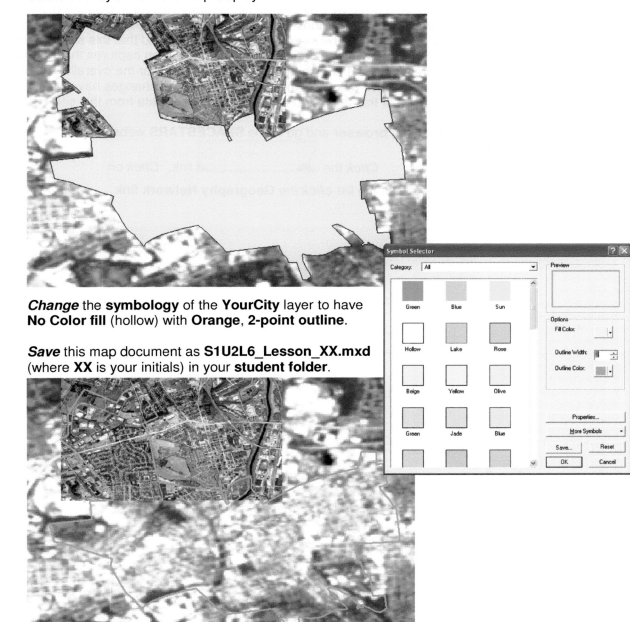

7.   ***Change*** the **symbology** of the **YourCity** layer to have **No Color fill** (hollow) with **Orange, 2-point outline**.

8.   ***Save*** this map document as **S1U2L6_Lesson_XX.mxd** (where **XX** is your initials) in your **student folder**.

Downloading Shapefile Data:
Up until this point, you have used data that has been provided for you. Many times, when you build a GIS project you have to find and acquire data that will be used in the project. This activity will lead you through the exploration of some commonly used online data sources. Keep in mind that data can come in a variety of formats – some data can be used immediately in a GIS software program. Other data must be manipulated and prepared for use. You will explore various types of data in this activity.

<u>Geography Network:</u>
Geography Network is a website that allows you to create online maps and download different types of data. This site is supported by Environmental Systems Research Institute (ESRI) the makers of ArcGIS software. The instructions for this activity provide detailed documentation about navigation of the Geography Network site. Please keep in mind that this site undergoes regular aesthetic changes and may appear differently than the screen captures that appear in this documentation. Rarely does the site change the names of links or the overall structure of the site. You may, however, have to look at the site more closely if changes have been made to the site since the publication of this text. You will download street data from this site.

9.   **Launch** your **Internet browser** and go to the **SPACESTARS** website at
        www.spacestars.org. **Click** the ⟨ **Links** ⟩ link. **Click** on
        Training Links . In the list **click** the **Geography Network link**
        www.geographynetwork.com.

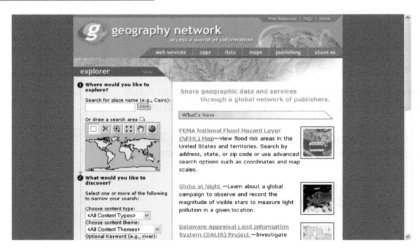

10.   **Scroll** to the bottom of the website and look at the various <u>Data</u> links that are listed.

11.   **Click** the • <u>Census TIGER/2000</u> link. A new window will open.

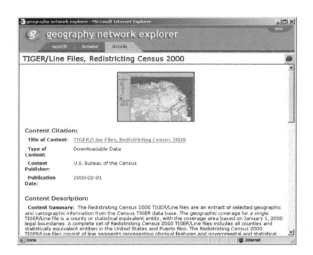

This page provides some information about the Census 2000 TIGER data.

12. **_Click_** the TIGER/Line Files, Redistricting Census 2000 link. Another window will open displaying a page on the ESRI corporate website.

13. **_Click_** the ▪ Preview and Download link in the leftmost column of the web page. Another web page will open.

14. Either **_select_** your **state** from the **drop down list** or **_click_** your **state** in the map image.

15. From the **County drop down list**, *select* **your county** and *click* [Submit Selection].
The available data layers for your county will appear in a list.

16. *Scroll* down the list to view the data that is available for your county.

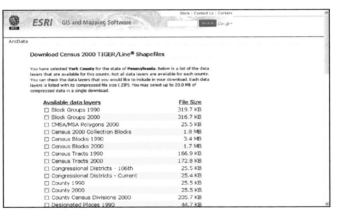

17. *Click* to *select* the ☑ Line Features - Roads data file.

18. *Scroll* to the bottom of the page and *click* [Proceed to Download]. You will then go to the download page.

19. *Click* the [Download File] button. The **File Download** dialog box will appear.

Click ___Save___ to save the data file on your computer.

In the **Save As** dialog box, ***navigate*** to your **student folder** and ***save*** this file. The file will be already named and will be in a zip format for faster download.

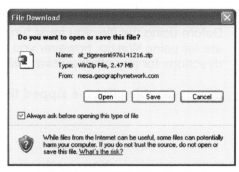

Click ___Save___. The download process may take a little time (depending on your connection speed).

20. When the download process is complete, ***click*** **Open Folder** to open your **student folder**.

21. ***Close*** the **Geography Network download window**.

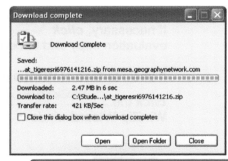

22. ***Scroll*** to find the **TIGER street network file** that you downloaded containing your county road data.

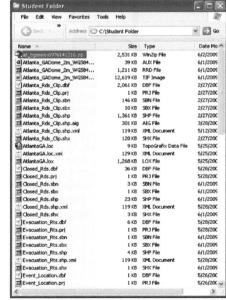

Extracting Zipped Data:
The file you just downloaded is a rather large file that was "zipped" to be downloaded quickly. Before using the file, you must "unzip" it to have access to its contents. The following directions are for using Winzip, however you may have another software that extracts zipped files. The directions for any other software will vary from these.

23.   ***Double click*** the **zipped file** to ***start*** **WinZip**

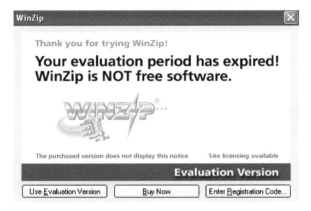

If necessary, ***click*** | Use Evaluation Version | to acknowledge that you are using an evaluation copy and to begin using **WinZip**. The WinZip interface will open.

If this interface of WinZip does not appear, ***click*** the | WinZip Classic | button on the **WinZip Wizard** window.

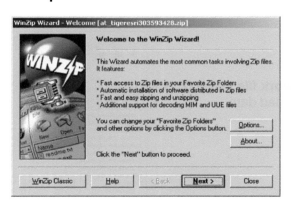

In the WinZip classic window, notice that there are two (2) files contained in the zipped data file. One is another zipped file that contains the data files that make up the street network data layer. The other is a Read Me html file that contains information about TIGER data files. (See the following graphic.)

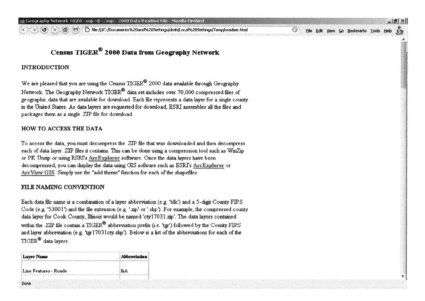

Double click the **lkA#####.zip** file (where **#####** is the TIGER code for your county) to start unzipping this file. If necessary, ***click*** [**Open**] to acknowledge that you want to open this file. The files that make up the street network shapefile are listed in the new WinZip window.

Even though you have downloaded a single data layer, there are several data files that make up a single shapefile data layer. There are a minimum of three (3) files that are necessary to create a shapefile: a main file (.shp), an index file (.shx) and a database file containing the attribute data (.dbf).

Click the [Extract] button.

In the **Extract** dialog box, ***navigate*** to your **student folder**.

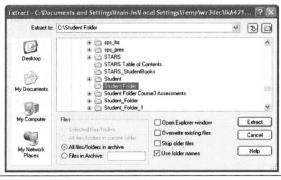

Click [Extract]. The files will be extracted and saved in your **student folder**.

---

24.   ***Close*** all of the **WinZip windows** that are open.

25.   Make **ArcMap** active again.

26.   ***Add*** 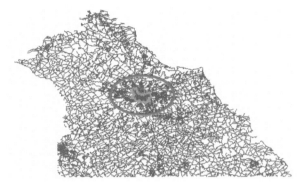 the **tgr#####lkA.shp** data layer that you just downloaded from your **student folder**.

Clipping a Data Layer:
The street network data layer that you just downloaded contains street segments for the entire county. You need only the streets in your city for this exercise. You will use ArcToolbox to clip the street network data layer using the city boundary data layer that you exported earlier in the lesson.

27.   ***Click*** the **Show/Hide ArcToolbox Window** button to display ArcToolbox tools.

28.   In the **Analysis Tools** toolbox, ***double click*** the **Extract** toolset to view the tools contained in it.

29.   ***Double click*** the **Clip** tool. The **Clip** dialog box will appear.

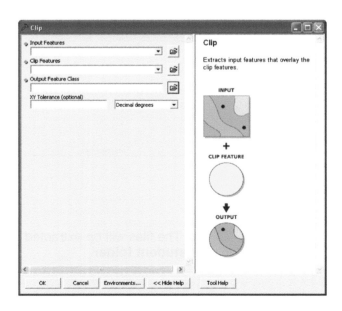

Terminology Tip

Clipping means using the features of one data layer as a "cookie cutter " to cut a piece of the features of another data layer. The input features to be clipped can be points, lines or polygons. The clip feature ("cookie cutter") must be a polygon.

Cluster Tolerance is the distance that determines the range in which features are made coincident. This means that in the clip shown that if the roads are clipped based on the city boundary with a tolerance of 10 meters, an additional 10 meter is taken outside of the boundary. If your GPS was known to be 10 meters off this compensates for precision

Select the **tgr#####lkA.shp** street network data layer as the **Input Features**. *Select* the **YourCity** layer as the **Clip Features**. (Notice that it is the only **Clip Features** layer available because it is the only polygon data layer that is included in the project so far.)

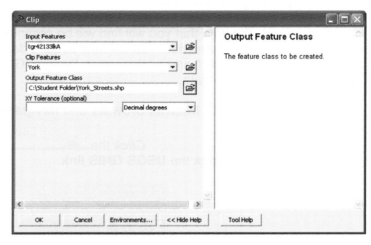

Save the **Output Feature Class** in your **student folder** as **YourCity_Streets.shp**.

Leave **Cluster Tolerance** blank.

Click  .

The **Clip** processing dialog box will appear and show the progress of the clip procedure.

When the processing is complete, *click* **Close** . The layer will appear in the Table of Contents and display in the map.

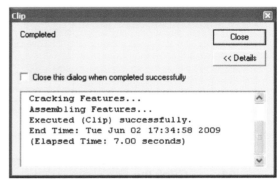

30. *Change* the **symbology** of the **YourCity_Streets** layer to **Orange** lines with **line width** of **1**.

   Close the **ArcToolbox** window.

   Turn off the street file for your **county** by unchecking it in the **Table of Contents**.

31. *Save* the project.

Downloading Tabular Data:
In some cases, the data that you will acquire for a GIS project will not be in a shapefile format. Sometimes, the data that you will find will be organized as a table and you will have to prepare the data for use in ArcGIS. You will now acquire data from the United States Geological Survey (USGS) Geographic Names Information System (GNIS) website. GNIS provides statewide geographic feature data.

32. **Launch** your **Internet browser** and **navigate** to the **SPACESTARS** website at www.spacestars.org. **Click** the **Links** link. **Click** on **Training Links**. In the list **click** the **USGS GNIS link** http://geonames.usgs.gov.

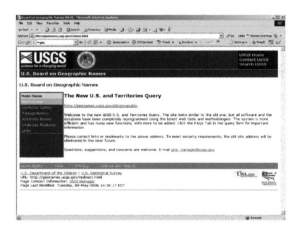

33. **Click** the **Domestic Names** link to download the data. When the list expands click **Download**.

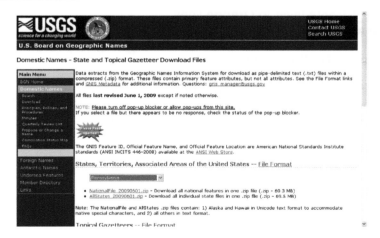

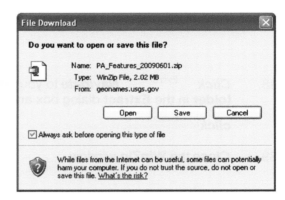

States, Territories, Associated Areas of the United States -- File Format

– Select state for download –

Select your **State** from the **pulldown list**.

34. *Click* ___Save___.

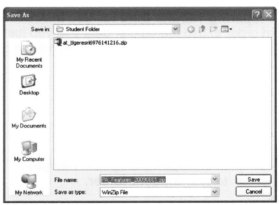

The **Save As** dialog box will appear. *Navigate* to your **student folder** and Save the file there.

To Unzip the Zipped Data:
If you downloaded the zipped data file, use the following instructions to unzip the data. If you did not download the zipped file, skip to instruction #40 to continue.

35. When the download process is complete, *minimize* the **Internet browser** and **ArcMap** windows on your computer screen.

36. Just as you did in the earlier download exercise, *navigate* to your **student folder** using either **My Computer** or **Windows Explorer**.

37. *Locate* the **ST_deci.zip** data file that you just downloaded and *double click* it to launch the

WinZip program. If necessary, *click* | I Agree | to acknowledge that you are using an evaluation copy and to begin using **WinZip**. The WinZip interface will open.

If this interface of WinZip does not appear, *click* the | WinZip Classic | button on the **WinZip Wizard** window.

The text table contained in the zipped file will be listed in the **WinZip** window.

38. *Click* Extract. *Navigate* to your **student folder** in the **Extract** dialog box and *click* | Extract |.

39. *Close* the **WinZip** window.

Preparing Tabular Data for Use in ArcGIS:
In order to use the tabular data that you just downloaded, you must prepare the data using a spreadsheet program. You will delimit the data so that it can be viewed as a spreadsheet and add field headers to the data.

40. *Launch* **Microsoft Excel**.

41. *Click* the **Open** button.

42. In the **Open** dialog box, *navigate* to your **student folder**.

   Change the **Files of type** setting to **Text Files (*.prn; *.txt; *.csv)**.

43. *Select* the **ST Features.txt** file that was downloaded. *Click* | Open |.

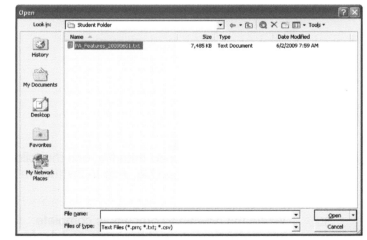

The **Text Import Wizard** dialog box will open.

The Text Wizard leads you through steps of organizing the text data so that it can be opened in Excel. The Text Wizard recognizes the format of the table as **delimited**, meaning that a character or tab is used to separate the fields.

Notice the **Preview** area of the **Text Import Wizard** dialog box shows how the data is organized. Notice that the fields are separated by the "|" character. (This character is located on the same key as the backward slash on the keyboard. You can type this key by pressing the Shift key and the \ key.)

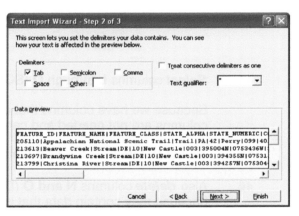

Click <u>Next ></u>. **Step 2** of the **Text Import Wizard** dialog box will appear.

The **Tab** delimiter option is automatically selected ☑ I̲ab . However, this is not the delimiter that is used for this data file.

Deselect the **Tab** delimiter option ☐ I̲ab .

Click the **Other** delimiter option ☑ O̲ther: .

In the box to the right of the ☑ O̲ther: delimiter option, you must type the | character. To do this, **press and hold** the **Shift** key and **press** the \ key.

When you do this, the **Data Preview** changes and the data contained in this file organizes based on the | delimiter.

At this point, **click** <u>Finish</u> to complete the text import process. The text data will import in spreadsheet format.

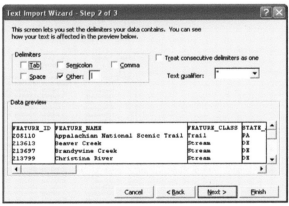

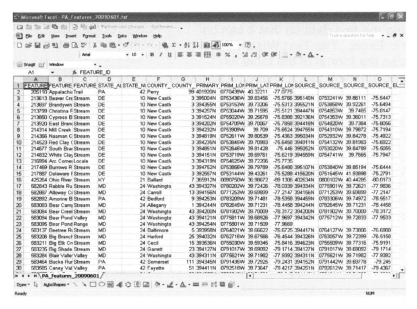

44. **Select** columns **L** through **O** (SOURCE_LAT_DMS through SOURCE_LONG_DEC).

   Because we have columns that already have the Primary Longitude and Latitude these columns are not needed and can therefore be deleted.

45. **Select** Delete from the **Edit** menu.

46. Also **delete** columns **N and O** (Date Created, and Date Edited) from the spreadsheet because they contain data that is not necessary for this study.

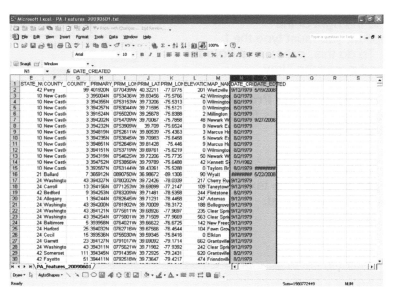

   ArcMap will not read field headings with spaces or special characters (punctuation marks, symbols, etc.) except underscores. Since no spaces are allowed, an underscore must be used to separate words. For example, "State Name" is not a valid header due to the space between state and name. It must be written as "State_Name". ArcMap is also only able to display field headings that have 10 characters or less. You will now edit the header row to the spreadsheet.

Enter the following **fieldnames** in the first row of the spreadsheet in the cells specified:

FID (Feature identification number, cell A1),

NAME (name of the feature, cell B1),

CLASS (classification of features, cell C1),

STATE (state abbreviation, cell D1),

STATE_FIPS (state FIPS numeric code, cell E1),

COUNTY (county name, cell F1),

COUNTY_FIP (county FIPS code, cell G1),

LAT_DMS (latitude in degrees minutes seconds, cell H1),

LONG_DMS (longitude in degrees minutes seconds, cell I1),

LAT_DD (latitude in decimal degrees, cell J1),

LONG_DD (longitude in decimal degrees, cell K1),

ELEVATION (elevation where feature is located, cell L1), and

MAP_NAME (name of the USGS topographic map that the feature would be located on, cell M1).

Remember the **Digital File Formats** information that you viewed in the **Read Me** file on the GNIS website. This is where that information is useful to determine exactly what data is contained in these columns.

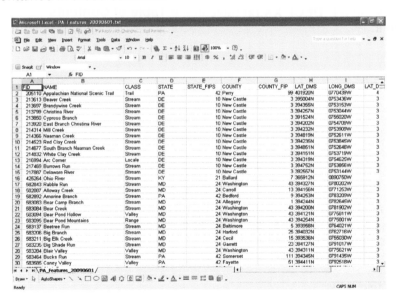

47. **Save** this spreadsheet as **ST_DECI_XX.txt** (where **XX** is your initials) in your **student folder**. The following dialog box will appear warning you of formatting incompatibilities:

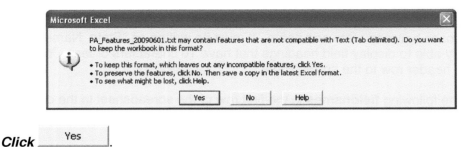

Click __Yes__.

48.   ***Exit* Microsoft Excel.** When prompted to save changes, ***click*** __No__.

Adding & Displaying Tabular Data to ArcMap:
Now that you have prepared the text data in Microsoft Excel, you can add the text table to ArcMap and then use display techniques to view the data in the map display.

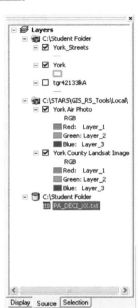

49.   Make **ArcMap** active again.

50.   ***Add*** the **ST_DECI_XX.txt** table from your **student folder** to the **ArcMap** project. When you add the table, the **Table of Contents** should switch to **Source** mode so that you can see that the table was added.

51.   ***Right click*** the **ST_DECI_XX.txt** table in the **Table of Contents** and ***open*** it.

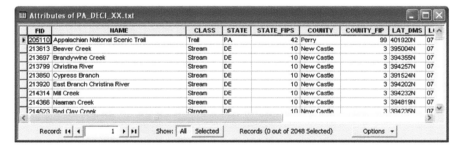

Notice that the fields are contained in the table as you prepared them in Excel. You probably also remember that longitude and latitude coordinates were given for each geographic feature contained in this table. You will now use those coordinates to map the data in the map display.

52.   ***Close*** the text table.

53.   To map the data, ***right click*** the **ST_DECI_XX.txt** table in the **Table of Contents** and ***select* Display XY Data...** from the **Context Menu**. The **Display XY Data** dialog box will appear.

The X and Y fields are already specified as **LONG_DD** and **LAT_DD** fields, respectively. If you think of lines of longitude and latitude as an XY grid, the lines of longitude (although they run north and south) measure east and west of the Prime Meridian (the 0 line of longitude) and would give the X value on the grid. Likewise, latitude runs east and west but measures north and south of the Equator, thus giving the Y value on an XY grid.

Although the **Display XY Data** dialog box provides an opportunity to specify spatial reference, you do not have to do this here. Once the data points are plotted and the features are exported as a new shapefile, you can specify spatial reference then.

To display this data, *click* **OK** .

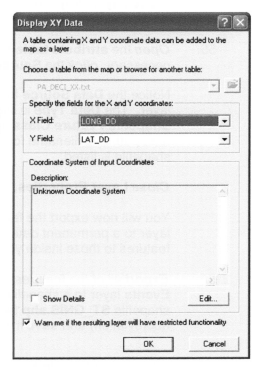

If the following message appears, *select* **OK** to allow it to create an object ID field.

Keep in mind that this table contained data for your whole state. You will use selection methods to identify only those features in your city. Notice that the features created from the table are displayed as an **Events** layer. An **Events** layer is different from shapefiles in that you cannot use an Events layer in analysis in ArcMap.

□ ☑ PA_DECI_XX.txt Events

54. To verify this, *open* the **Properties** for the **Events** layer. *Click* the **Source** tab.

Notice that the **Data Type** is **XY Data Source** with the **ST_DECI.XX.txt** file set as the table.

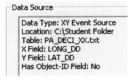

Data Source

Data Type: XY Event Source
Location: C:\Student Folder
Table: PA_DECI_XX.txt
X Field: LONG_DD
Y Field: LAT_DD
Has Object-ID Field: No

For **Events** layers, no spatial data file is created. The locations are simply displayed within the ArcMap map document using the coordinate table found in the text table.

***Close* Layer Properties** for the **Events** layer.

55. ***Open* the attribute table** for one of the other feature layers in the **Table of Contents**. If necessary, ***click* the Source** tab.

Notice the **Data Source** for this layer. The **Data Type** is listed as **Shapefile Feature Class** and the data path and filename for the shapefile is listed.

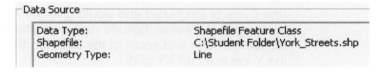

Data Source
Data Type:	Shapefile Feature Class
Shapefile:	C:\Student Folder\York_Streets.shp
Geometry Type:	Line

***Close* Layer Properties**.

You will now export the features in the **Events** layer to a permanent data layer before you clip the features to those inside your city.

56. ***Export*** all of the features in the **ST_DECI_XX.txt Events** layer to a shapefile. ***Name*** the new shapefile **ST_GNIS.shp** (where **ST** is your state abbreviation) and ***save*** it in your **student folder**.

When prompted, ***click*** Yes to add the new data layer to the map.

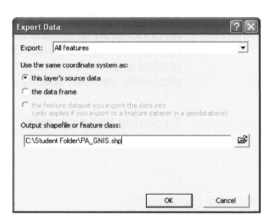

Export Data

Export: All features

Use the same coordinate system as:
• this layer's source data
○ the data frame
○ the feature dataset you export the data into
(only applies if you export to a feature dataset in a geodatabase)

Output shapefile or feature class:
C:\Student Folder\PA_GNIS.shp

OK   Cancel

57. ***Switch*** the **Table of Contents** back to **Display Mode**. ***Remove*** the **ST_DECI_XX.txt Events** layer that you created earlier.

58. ***Click*** to view **ArcToolbox** tools.

In the **Analysis Tools** toolbox in the **Extract** tool set, ***clip*** the **ST_GNIS.shp** layer (**Input Features**) using your **city polygon** layer (**Clip Features**) and ***save*** the **Output Feature Class** as **YourCity_GNIS.shp** in your **student folder**. Leave **Cluster Tolerance** blank.

Click OK to perform the clip.

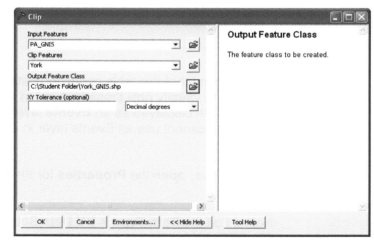

Clip

Input Features
PA_GNIS
Clip Features
York
Output Feature Class
C:\Student Folder\York_GNIS.shp
XY Tolerance (optional)
Decimal degrees

Output Feature Class

The feature class to be created.

OK   Cancel   Environments...   << Hide Help   Tool Help

59.   When the process is complete, *remove* the **ST_GNIS.shp** data layer from the **Table of Contents**.

       Close the **ArcToolbox** window.

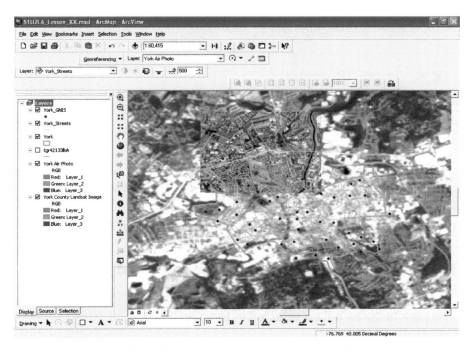

60.   *Open* the **attribute table** for the **YourCity_GNIS.shp** layer that was just added to the map display.

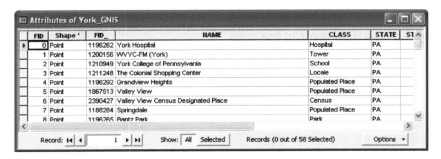

       You will need to select the features of each class in this data layer and export them as a separate data layer.

61.   *Close* the **attribute table**.

62.   Use **Select by Attributes** to highlight all of the **schools** included in the **YourCity_GNIS.shp** data layer.

       The school features will be selected in the map display.

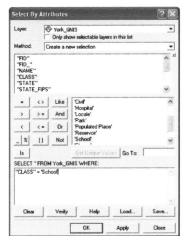

63. ***Export*** the **selected features** from the **You**...

When prompted, ***add*** the new data layer to the map.

64. ***Click*** the **Clear Selected Features button**  on the Tools toolbar.

Build a new query to highlight all of the **parks** included in the **YourCity_GNIS.shp** data layer.

The park features will be selected in the map display.

65. ***Export*** the **selected features** from the **YourCity_GNIS** data layer as a new shapefile. ***Name*** the new layer **YourCity_parks.shp** and ***save*** it in your **student folder**.

When prompted, ***add*** the new data layer to the map.